Legends *of the* Common Stream

Other Books from Bright Leaf

House Stories
The Meanings of Home in a New England Town
BETH LUEY

Bricklayer Bill
The Untold Story of the Workingman's Boston Marathon
PATRICK L. KENNEDY AND LAWRENCE W. KENNEDY

Concrete Changes
Architecture, Politics, and the Design of Boston City Hall
BRIAN M. SIRMAN

Williamstown and Williams College
Explorations in Local History
DUSTIN GRIFFIN

Massachusetts Treasures
A Guide to Marvelous, Must-See Museums
CHUCK D'IMPERIO

Boston's Twentieth-Century Bicycling Renaissance
Cultural Change on Two Wheels
LORENZ J. FINISON

Went to the Devil
A Yankee Whaler in the Slave Trade
ANTHONY J. CONNORS

At Home
Historic Houses of Eastern Massachusetts
BETH LUEY

Black Lives, Native Lands, White Worlds
A History of Slavery in New England
JARED ROSS HARDESTY

At Home
Historic Houses of Central and Western Massachusetts
BETH LUEY

Flight Calls
Exploring Massachusetts through Birds
JOHN R. NELSON

Lost Wonderland
The Brief and Brilliant Life of Boston's Million Dollar Amusement Park
STEPHEN R. WILK

Legends *of the* Common Stream

John
Hanson
Mitchell

BRIGHT LEAF
Amherst and Boston
An imprint of University of Massachusetts Press

I Believe I'll Go Back Home has been supported by the Regional Books Fund, established by donors in 2019 to support the University of Massachusetts Press's Bright Leaf imprint.

Bright Leaf, an imprint of the University of Massachusetts Press, publishes accessible and entertaining books about New England. Highlighting the history, culture, diversity, and environment of the region, Bright Leaf offers readers the tools and inspiration to explore its landmarks and traditions, famous personalities, and distinctive flora and fauna.

ISBN 978-1-62534-581-3 (paper); 582-0 (hardcover)

Designed by Sally Nichols
Set in Adobe Garamond Pro
Printed and bound by Books International, Inc.

Cover design by John Barnett, 4Eyes Design, Inc.
Cover art by Jill Brown, *Beaver Brook,* © October 2020,
courtesy of the artist.

Library of Congress Cataloging-in-Publication Data
A catalog record for this book is available from the Library of Congress.

British Library Cataloguing-in-Publication Data
A catalog record for this book is available from the British Library.

Scratchboard illustrations by Tamsin Barnes © 2020.
Courtesy of the illustrator.

For my parents, who got me started on all this

Time is but the stream I go afishing in
—Henry Thoreau

Contents

Preface

Many of the ancient villages and towns of Europe were located along streams and brooks and narrow rivers. These were colloquially known as the common stream, and in the Middle Ages most of the life of the village utilized the running waters of these brooks for watering livestock, or for ducks and geese, laundry, and, later in history, small water mills.

In the distant past, even before the Bronze Age, streams of this sort, as well as springs and natural pools, were populated by a host of water spirits such as Naiads and Undines, and also among certain cultures, evil, dangerous beings such as trolls, water goblins, or the amphibious monster, Grendel, who lived in a cave at the bottom of a deep lake.

These springs and streams were once revered and honored, either through regular offerings, or rituals, or the construction of shrines. Then, with the coming of Christianity into northern Europe, the spirits of the place were often transformed into Christian saints, and the springs and streams continued to be honored with shrines. The French cathedrals of both Chartres and Notre Dame in Paris were deliberately located on the site of ancient sacred springs, for example.

There is a wide, slow-flowing stream called Beaver Brook about thirty-five miles west of Boston that borders the eastern boundary of a tract of agricultural land known as Scratch Flat. This brook was an important resource for the native people who lived along its shores over the past ten thousand years, and until recently, it was also important to the English colonial farmers and their descendants who settled here in the late seventeenth century. Both these cultural groups lived in a world in which ghosts and water spirits, faeries, and demons of one variety or another were very real. Even the Christianity of the English did not obliterate acceptance of these beings. The Puritan fathers believed that forests along the brook were populated with the ancient demons of the pagan past, now transformed into agents of the Devil. There were screeching, fiery worms that flew by night over Scratch Flat; there were demonic black cats, and mischievous faeries and also bats and witches, one of whom supposedly haunted the Scratch Flat fields as late as 1722.

Beaver Brook lies just east of my house, and over a period of years, I got in the habit of walking down through a wooded cart track that leads to an open bank above the stream to sit in the morning sun and see what I could see from that singular point of view. As a result, thinking back on the former residents of the watershed and their different cultural attitudes toward the nature of the place, I began to think of Beaver Brook as a common stream, although it was generally less populated and wilder than the village streams of Europe.

It came to me one day, thinking (as usual) about nature and history and the fate of the natural world in the face of human events, that the whole story of the interrelationship between human cultures and the natural world could be summarized in the history and the related myths and lore of the plants and animals that could be found along this singular stream.

This was actually no idle entertainment. The fact is, the rich body of myth, folklore, and legend have been all but deserted and forgotten in our time. This may seem logical in an age in which scientific discovery has explained most of the phenomena that so mystified the ancients. But our contemporary alienation from myth separates us not only from nature, but also from enduring lessons on human existence. Myth was once indispensable to cultures around the world. Belief in traditional mythologies and legends helped illuminate the mysteries of our place in the world; they answered the seemingly unanswerable questions of nature: the return of birds and vegetation after the silence of winter, the cycles of the sun, moon, and stars, birth, and death, and also, and not incidentally, what happens after we die.

Down by the brook, after much consideration of our contemporary separation from the world of nature, I decided to expand the story of the natural history of my literal common stream to include the universal common stream of the collective myths and legends of worldwide cultures. This book is the result of that research.

Legends *of the* Common Stream

CHAPTER ONE

THE DEER MOON

I know a bank above a quiet stream where the heron and the bittern haunt the wide marshes and dark waters slip between high walls of rushes and reeds. Just before dawn on Christmas morning a few years ago I went down to the stream bank to watch the day unfold. It was a day not unlike that of other days that I visit the stream bank except that this was the official beginning of the winter season. The sky was cloudy, and the air was filled with that close, watery scent of coming snow. I went out anyway and rather than following the unused cart track that leads through the woods to the stream, I took a short cut and clambered through the undergrowth that now covers what were once the hayfields that rolled down to the brook in a series of terraces.

Bushwhacking in this manner through tangles of bittersweet and brambles and skirting walls and blow downs and entering, finally, into a somber dark forest, with only a slight suspension of disbelief, you might think for a second or two that you had entered into a timeless wilderness here. But in fact, this is a fast suburbanizing region of former farmlands and apple orchards, and in order to live with the myth of the past, you have to go burrowing into the night woods and seek out those few isolated spots in the area, such as the stream bank.

Down at the brook that morning, the grass on the bank was still snow free and I sat cross-legged above the waters and waited for sunrise. There was nothing to wait for, actually; this day was characterized by that flat, seamless pall of sky that prefigures snow. A few flakes began drifting down as I watched, followed by a few more and then by a steady but light fall. As I waited, a pale, faded image of the white winter sun, much filtered through the mists and drifting flakes, showed itself—a silvery coin behind the black branches of the trees. It was an odd, somehow portentous sunrise, the type of light into which the native shamans and Puritan ministers of these parts might read dark omens.

Among Christian cultures, Christmas is the day when it is believed that a savior of the world was born, a divine figure who would mark the beginning of a golden age of peace and glory. But shortly before the historical birth of Christ the same date was celebrated by a popular Roman cult whose chief luminary was Mithra, who was not exactly a god, but a representative of the true and only god—the sun. Long before that, by a thousand years or more, the day was celebrated as a solstice festival. This was, for all the cultures of the world, an absolutely critical date. In earlier times, before the advent of organized religions, the ancients watched the fading light and shortening days with dread. It was not clear that the sun would not carry on in its decline and never return. To halt this process, shamans and holy men would carry out rituals, most of them involving fire—the thought being that the fiery sun would be nourished by earthly light. Some of the local Native American tribes would shoot burning arrows into the night sky at this time of year, and around the world bonfires were lit and processions of torch-bearing celebrants circled the fire-chanting ritual phrases to bring back the sun.

In ancient Egypt, pharaohs, who were considered the earthly embodiment of the sun, used to perambulate the temple walls to encourage the real sun in its daily course. And each year at Rhodes, the ancient Greeks would cast a chariot and four horses into the sea to refresh the worn-out team of Helios, who drove the chariot of the sun across the sky by day and sank in the western seas at dusk.

Here, in the silence of the drifting snow, above the quiet flow of black waters below the stream bank where I was sitting, you would know nothing of this. On such a morning, the peace of wild things descends, time stalls out, and even runs backward on certain days so that I might be sitting here four hundred years ago when, not far from this bank, a tractable old Pawtucket man named Tom Doublet tended his fish weir. He had worked as a peace negotiator during the local King Philip's War in 1675 and, after the war moved his wickiup to the banks of the brook where he lived out his days, presumably reflecting on the vast cultural changes that had taken place in the course of his lifetime.

I suppose in some ways I was doing the same thing here, shuffling through my days, planting trees on my property as a hedge against disaster, a statement of faith in a future and the abiding force of nature to recover itself. And every day, as a way of setting things straight, I would come down to the bank above the slow stream where the water lilies bloom, and frogs call from the green tangle of vegetation that lines the stream.

The site offers a good vantage point for viewing the surrounding landscape, and if you are of the type that appreciates the implications of the flow of dark waters from source to sea, and the unfolding progress of the seasons, it's one of the only places around here where you can gain some perspective on the regular cycles of the natural world—as opposed to the alarming events of current history.

I've been coming to this spot almost every day for over twenty-five years now. I come in winter, when the waters are ice-bound and laced with the tracks of otters, foxes, and coyotes. I come in spring, when the stream is in full flood and the cries of red-winged blackbirds fill the air and the green spears of cattails are just pointing up. I'm there in summer, when the water lilies bloom and turtles bask along the shores, and I come in autumn, when the marshes turn lion brown and flights of wood ducks crisscross the open sky. It is a good place to think. A good place to be alone. In the twenty-five years that I have been coming here I have only regularly seen one other human being on the marshes, an older self-taught nature man named Jon Holbrook who commonly patrols the brook in his canoe—standing up no less—so as to see better over the tall cattails.

And yet, in a larger sense, I am never entirely alone in this place. Although I rarely see them, there is an active lodge of beavers not one hundred yards downstream. Muskrats are common here, and there is a resident otter family somewhere near this spot, whose signs I spot every day. A fox passes by almost daily, coyotes patrol the banks, and the wide freshwater marshes of the stream offer one of the richest areas for bird life in the region. Herons feed along the banks, and in the green depths of tall rushes and reeds there is an active population of long-billed marsh wrens. In winter, the little golden-crowned kinglets flit in the clumps of dried cattails and buttonbush, winter wrens dart in the underbrush, and in spring the red-wings, yellowthroats, and wrens set up a veritable city of birds. Huge undulating water snakes wind through the dark waters in summer, hunting the green frogs and bullfrogs that sound out from the grassy interiors all season long, and in autumn and spring the marshes are a stopover for migrating ducks and geese and hooded mergansers.

The brook itself is part of the Merrimack River watershed and rises in the uplands and swamps and sinks about ten miles upstream from my bankside sanctuary. The narrow stream gathers its waters from swampy springs and seeps, and then widens and deepens, and for the next nine miles or so runs between banks that are wild and wooded, save for the inroads of a few houses and farms.

About a half mile upstream from my post on the bank, it crosses under a few roads that were originally Indian trails, runs past the ruins of a nineteenth-century sawmill, and then dives beneath a bridge that carries a veritable tidal stream of cars and trucks in a huge ring highway that holds in Boston. Where it passes in front of the grassy bank it runs through a spot which for untold centuries was the site of a fish weir maintained by a small band of either Nipmuck or Pawtucket Indians, the last of whom was the local historical figure named Tom Doublet.

Below the bank, the marshy floodplains are wide, almost like a prairie, and the stream wanders slowly through these rich grasses, around oxbows and meanders and bends. If you paddle this part of the brook you have to travel nearly three miles to gain but one mile as the crow flies. Once through the wide marshes, the stream narrows again, then crosses under a road and runs into a shallow lake, surrounded by housing where a former gristmill was located.

After 1800, as the industrial revolution progressed, the little grinding mill was converted to a small woolen mill, which itself was converted into an industrial building powered by electricity in the 1930s and ultimately, in our time, abandoned for a while, and then transformed into condominiums and offices.

This obscure little falls has an interesting, perhaps telling, prehistory. The people who lived in this region over the last two or three thousand years were members of a generally peaceable group of Native Americans. They were surrounded to the east and northwest, however, by notoriously warlike tribes, the Massachusetts Indians to the east, and the dreaded Mohawks to the northwest. In their time, the Massachusetts Indians, the same tribe that first encountered the Puritans who settled Boston, were a powerful entity, controlling a huge section of the lands above the North River in southeastern Massachusetts, all the way up to what is now Maine and inland to Concord and beyond. The Iroquois nations, in upstate New York, were even more violent and would carry out seasonal raids on the more generally peaceful coastal tribes. They were known at the time as the Maquas, or man-eaters, and were feared for their warlike demeanor. Tom Doublet's father was killed in one of these seasonal raids as he was tending his fish weir on the brook just south of the high bank where I spend my mornings.

This was not, in other words, an entirely peaceful place in the past. However, in what is known as the pre-contact era, that is, the time before the coming of the European people, the low waterfall here was an important fishing spot for the native peoples. Salmon, eels, alewives, and other herring species would migrate inland from the sea and collect below the falls in spring. Given the prevalence of wars and raids among the native people, it was not uncommon to have some local fighting going on after the retreat of the winter snows. And yet, as soon as the fish would run, word would spread and the warring bands would call a truce and gather at the falls to collect fish.

Below the dam that has now replaced the waterfall, the stream widens and runs deeper and eventually empties into the Merrimack River, not far from a trading post where the invading English would often meet with their sometime enemy Indians to trade goods. From that point the Merrimack carries on in a northeasterly course through rapids and falls, and eventually enters into Massachusetts Bay where the town of Newburyport is now located.

. . .

Snow was falling heavier now, obliterating my tracks and filtering the world into a monochrome plate where a few dark bare branches formed an unreadable calligraphy. But it was a light airy snow, and occasionally the sun would pierce the fog and send shooting rays across the tawny grasses of the marsh. It reminded me of medieval tapestries I had seen in the Cluny Museum in Paris, depicting the imaginary unicorn, which, incidentally, was traditionally associated with the sun in pre-Christian cultures. It occurred to me that I had perhaps been here longer than I realized; snow had collected on my shoulders, so I got up to leave.

I presumed I would be alone out here on this of all days, but on my way home as I was walking up the old cart road that leads to my house, I was surprised to meet a deer hunter, just setting out. He was a round-faced man of about thirty, with black hair, cut in bangs, and round, slightly Asian epicanthic black eyes. In fact, he looked a little like one of the ancient Siberian mammoth hunters, the progenitors of the American Indians I had been thinking about earlier by the brook. But he spoke perfectly good English, albeit with a strong local accent. We chatted for a while about the resident deer populations and the changing patterns of the wanderings of the herds to and from the brook and whatever other common ground we two could be comfortable with. He eyed me suspiciously all the while, though—hunters in these suburbanizing woodlands are not always favored, and he knew it. Furthermore, I knew that he knew that I was trespassing on what amounted to "his" land—or at least his territory.

This track leads down to the brook through a local sportsmen's club that is theoretically open only to members. Signs are posted all around the twenty-seven acre plot of land warning away trespassers. But primal animal that I am, I have never paid any attention to these warnings, unless there were cars in the small parking lot. He knew I was the transgressor here, but he said nothing, so I said nothing.

I actually appreciate hunters in spite of the fact that I am not one of them. They are remnants of an earlier time and serve, if nothing else, to encourage me in my fantasy that the past, with its ancient weight of animal and human associations, still endures in the present, even though we don't always recognize it. Hunters often know the lay of the land better than the average local citizen. Furthermore, perhaps ironically, it turns out that some of the most peaceful social groups in the world are traditional hunting societies, such as the Ba'mbuti pygmies of the Ituri Forest, in West Africa.

My conversation with the local hunter that morning was made all the more tense inasmuch as this man was armed with a mean-looking shotgun that he held crooked in his right elbow. If I have my facts straight, from the little I know of these matters, this looked to me like one of those guns that some deer hunters carry that allow you to fire off a large number of rounds between reloads. I did not dare ask him about his weapon, however, for fear of invoking in him a suspicion of anti-gun sentiments. I knew from the club literature that one of the group's primary mandates was the protection of Second Amendment rights.

Interestingly enough, however, I knew also from their founding documents that their twenty-seven acres of land were originally set aside to protect wildlife. So arms notwithstanding, the hunter and I shared certain values.

I was still nervous about his gun, though, and tried my best to keep my eyes off it and avoid any discussion of why it is necessary to carry a powerful weapon that can fire off a number of rounds in a very short time into the body of a deer when one well-placed shot would do. Had we come to this, I might have pointed out to him that the hunters who lived along this stream bank for most of human history, that is to say Tom Doublet and his band, were able to bring down deer and even woolly mammoths with spears, and later, after the bow was developed, with a single arrow shot. I might also ask him if, after he had slain his deer, had he lain its head on his lap and stroked the deer's nose three times, asking for forgiveness and thanking it for giving up its life, as certain native tribes in these parts used to do.

As things turned out, however, the two of us parted amicably, he off to the forest to kill deer, and I home to the bosom of my family and a warm Christmas dinner.

CHAPTER TWO
THE HUNGER MOON

Shortly after Christmas that year, as is often the case, a dramatic cold spell swept in and the waters of the brook froze abruptly. There was no heavy snow, however, and after a few bitter nights, the brook was frozen hard enough to skate on. I decided to skate the length of "my" section of the brook, which runs from a bridge over the old Great Road to the south and ends at a bridge just east of a lake known as Forge Pond. It's a winding course of three miles or so, running through high walls of cattails for most of the route.

The stems of the cattails are beaten down by wind and snow by mid-winter, but early in the season, before the first heavy snows, you glide along between high, brown walls, curving east, then west, and trending all the while northward. In summer when the cattails are at their height, this same passage can be a voyage through a green hell, with side channels that look for all the world like the mainstream brook leading you off to dead ends, as well as huge looping oxbows such that at some points you find yourself paddling due east or west or even southward—the wrong direction.

Along the brook after each snow, the non-hibernating animals make themselves apparent. On the wooded banks, you can see tracks of grouse, turkey, and the scatterings of fur or feathers where owls and rough-legged hawks have struck their prey. And on the brook itself, after the first snows, the long furrows of skidding otters often lay a single track that disappears suddenly into open pools of dark water where springs have kept the stream ice-free. Flocks of red-polls, crossbills, and pine grosbeaks pass over the brook from time to time, and on warm days, during the January thaw, around the twelfth to the fifteenth of the month usually, you can see soldier beetles on south-facing tree trunks and the normally aquatic stoneflies, resting on rocks above the stream. Certain species of insects, such as the wood cockroach, a native wild species, seek shelter in the fluff of frayed cattail heads, and later, on warm days in February, I sometimes see mourning cloak butterflies along the wooded shores. They hibernate in winter and come out on the first warming days of the season.

That day, January 6, I got dropped at the bridge on the main road, tied on my skates, and set out. This was smooth sailing for once; the ice was a black ribbon curving around the turns and through straightways, and I made good time. In fact, I would have gotten to the downstream bridge in a half-hour or so except that I had to stop often to look at bird nests and explore the side channels to examine muskrat lodges and also a big beaver lodge about halfway between the bridges. This I had to clamber over—no easy task in any season, much less in skates.

January was the traditional Hunger Moon of the Eastern Woodland Indians, the outright depth of activity in the natural world. The restless bands of Siberian hunter-gatherers who crossed over the Bering Strait land bridge about 15,000 years ago were no strangers to these unforgiving weather conditions and had adapted to their environment. The people who eventually settled in this region, the Eastern Woodland tribes, survived on what they had stored up in corn and dried berries, and venison. They would mix all this together with bear fat and live on a sort of trail food they called pemmican. But when conditions were especially bad, certain older members of the tribe might choose to leave the encampments and walk off alone and allow themselves to die of exposure.

By 1620, when the Pilgrims arrived at Plymouth, that three- to four-thousand-year period of the culture of Native Americans was coming to an end. Plagues of as yet unidentified diseases had wiped out nearly ninety percent of the population, and by the late seventeenth century, wars with the English and acculturation were finishing them off.

Tom Doublet, who maintained his last residence just upstream from my little sanctuary, was one of the last surviving members of his tribe in the immediate area, and he himself had been much influenced by the culture of the invading English. He was a member of a mixed group of local Indians who had been converted to Christianity in the 1650s and lived in the nearby village of so-called Praying Indians known as Nashoba Plantation that was founded by the Puritan minister John Eliot in 1654. Doublet spoke English, and although he still knew all the traditional survival techniques, he essentially lived between two worlds, half-Indian, half-English.

About a mile or so down from the first bridge, I passed under the bank of my sanctuary, and inasmuch as the ground was still snow free, I climbed out and sat for a while to rest in the late morning sun. Sheltered as I was from the northwest wind by the forest wall behind me, the place still had a little warmth to it and I sat listening to the sounds of the floodplain: the periodic whisper of the wind, the creak of freezing ice, and a single chilled chickadee call. All was stillness and peace, and then, in the midst of this quiet, I heard another singular bird call, an elaborate, bubbling, and intricate song of the winter wren, a species of wren that nests in Canada and moves "south" (as if New England could be called "south") to the wooded banks and brushy tangles around Beaver Brook.

It was not far from this spot that Tom Doublet's father maintained his fish weir. He was killed while tending his weir by raiding Mohawks. Tom himself was born around 1620, about ten years before the Puritans settled on the little spit of land that later became known as Boston. By the mid-1640s, when Doublet came of age, the English had successfully broken up many of the tribal loyalties, and by 1654 had even managed to convert certain groups to Christianity, Tom Doublet among them.

Documents of the late seventeenth century characterize Doublet as a "tractable Indian," although that too may be a reflection of the man as the later chroniclers remembered him. He was old by then and lived a solitary life in his wickiup by the brook. As far as we can tell, reading between the historical lines, Doublet never did go to war. He had joined a religion that preached peace as a doctrine but then made war on his people. He did his best to negotiate a peaceful resolution between these two notoriously violent cultures, and then, having failed, retired to count his remaining days on the banks of Beaver Brook, perhaps embittered, perhaps taking solace in the dark passage of the waters and the sweep of the seasons.

In spite of his Christianity, however, it is likely that he, like so many others in his Christian Indian village, maintained his old traditions, albeit secretly. After he killed a deer, he would likely have stroked her nose three times and asked her forgiveness, or thanked her for giving her flesh to him. He would have believed that the bear, which was a common resident of the Beaver Brook watershed in his time, had certain human qualities and was due respect in both life and death. Either way, with his old traditions and his new religion he must have lived in a certain comfortable harmony with the natural world, perhaps soothed by two sets of myths he used to sustain himself.

Following this little morning sojourn on the bank, I skated out onto the ice, moved slowly downstream along the bank, and saw something dark flitting around in the buttonbushes. I never did get a good view. But the song was a reminder of the date. This happened to be January 6, an important date in various traditional European cultures, and also the time of a celebration in the British Isles known as Wren Day. The holiday is something of a moveable feast, usually celebrated on Saint Stephen's Day, shortly after Christmas, but it also used to be played out in early January in some areas.

On the given day, "wrenboys" would set off and hunt the lanes and hedgerows to catch wrens, which they would then tie to poles and carry through the villages, singing a traditional wrensong and soliciting donations. The holiday died out in England in the mid-twentieth century for the most part, but I used to know an Irish woman who told me that when she was young, she was a wren girl. "It was very like your trick-or-treat Halloween holiday," she told me, "only by my time, we no longer captured wrens; we had a little model bird we would tie to the pole."

The better-known holiday on this date is Twelfth Night, which in Christian Europe was a traditional celebration that marked the end of the twelve days of Christmas and the beginning of winter. On this night, among other myths, the animals were supposedly able to talk. I used to tell my two children

that our dog and two cats were able to r
midnight on the sixth, and since they we
by that time, they believed me. They as]
dog said. "Nothing interesting," I explai...
`Feed me. Cookie? Walk?'

"Also, 'I'll love you forever!'"

Wassail was consumed in quantity on the Eve of January 6, and in parts of southern England generous libations were offered to the fruit trees to increase fertility. In France, the date marked an older celebration, a sort of All Fools' Day when the world was turned upside down. Priests became commoners, a peasant was crowned king and Lord of Misrule, and the night was filled with music and dance and what Puritan critics described as lascivious mischief. A cake was prepared embedded with a bean, and whoever got the bean was declared king (or queen—there was a separate cake for women). The two were crowned and "married," honored with unholy toasts and accompanied by music and dance.

In the late Middle Ages, the holiday was declared a pagan rite and was banned by the Church, but that did not manage to halt the celebrations.

My source on much of this European lore was a nature man named Jon Holbrook, better known by his middle name, Riggs, to his friends. Riggs was a professorial old man, probably in his late seventies or so when I first met him, and he had lived a varied life of many careers, one of which was a sojourn in academia. He had taught English as an adjunct professor at a number of colleges and schools but had never settled at any one. He had, I was later told, inherited some money in mid-life and deserted teaching altogether in order to become a perpetual student so as to follow up on a host of long-standing interests, one of which involved an archeological dig in Guatemala. He told me once that he had spent half his

CHAPTER TWO

.e reading stray leaves from strange literatures, and the other half wandering in strange cultures.

I learned later from his younger wife, Rosemarie, that he was born in London and had been studying Italian in Florence when the Second World War broke out. He wanted to come home to join the army, but when the SOE, the secret service agency, learned that he was fluent in Italian and a student, they told him stay in Florence and keep his eyes and ears open. His wanderings and spotty academic career took shape when he came to visit America.

Riggs was an older member of a roughly associated group of people who lived along the western banks of Beaver Brook. There was, among various families who lived there at that time, a plant collector and local conservationist, an Italian gardener, another gardener, a single woman, and me and my somewhat eccentric extended family and an assortment of various dogs and cats. I had lived on Scratch Flat with my first wife for a number of years, and then we split up and divided the four-acre property and I eventually built a house and created an extensive garden, based roughly on a Renaissance Italian design. I married again, and my current wife had two children, which increased the members of the younger generation to four on what became known as "the compound." My daughter from my first marriage moved onto the property next door with her husband, had two more children, and thereby increased the number of children who grew up exploring the local woods and the brook. It was the children of these marriages who revived my interest in the old nature legends and myths, which I tried to relate to them on our various nature expeditions around the local woods and fields. No that any of them listened or even cared in this fast-moving electronic age, but at least I tried.

I had first met Riggs Holbrook out in the marshes of the stream. He was coming upstream as I was drifting downstream,

engaged in essentially the same thing that he was doing—
looking for wildlife. He had an odd way of exploring; he had
a long-handled paddle and would commonly stand in the
stern of the canoe, binoculars suspended around his neck. It
is little wonder that when I first saw him, I slowed my boat
and engaged him in conversation.

One of the things I first noticed was that he looked a lot like
photographs I had seen of William Brewster, the well-known,
late-nineteenth century ornithologist at Harvard, who also
used to search for birds on the marshes of the Concord River
in his old Rushton canoe. Riggs had dark eyes and a grey beard,
and he often wore an old slouch hat, the brim turned up above
his forehead, as did William Brewster.

Later I got in the habit of visiting him to find out what sort of
birds he was seeing. He lived above the Beaver Brook marshes in
an old hunting lodge that was built for a local hunt club in the
1920s. It was a dark shingle dwelling with a wide porch overlooking the marshes on the northwest side of the shores, where the
stream makes a wide curve before narrowing and making a dash
beneath the bridge and emptying into the lake.

We often spent time talking about birds and the local wildlife. But as I got to know him, our conversation strayed farther
afield. While he was teaching, he had posts in English-speaking
schools in various countries around the world—in France,
where he taught at the university in Aix en Provence, at the
American Academy in Rome, and in Mexico, and in Colombia,
where he first developed an interest in birds.

He had moved to the Beaver Brook area a few years earlier
and still had many friends in Cambridge, and he used to hold
somewhat bizarre parties, to which, as a sympathetic neighbor,
I was often invited. They would sometimes hold Twelfth Night
events at his house to which all his eccentric Cambridge friends
were invited, as well as a few chosen locals.

Following the traditions of the British Isles, they would serve wassail, and at midnight stream out to a small orchard patch in his backyard and honor the old apple trees. Rosemarie would bake the traditional French Twelfth Night cakes into which she placed small figurines of a king and a queen. Whoever got the figures in their serving was crowned.

Many of these traditional antics predate the advent of Christianity in Europe. In particular, there was the Roman holiday, Saturnalia, during which laws were suspended and there were festivities and celebrations, many of them involving fire and candles. Saturn was the god of agriculture and also libation and dissolution, and the carnival was associated with the harvest and also the return of light to the earth. The holiday was later subsumed by the Church and was said to be the day that the Three Kings arrived in Bethlehem, bearing gifts. In some Latin countries, Latin American cultures in particular, Christmas is still a twelve-day Saturnalia-like festival, known as the Festival of the Three Kings.

None of these holidays would have been appreciated by the English, who were living along Beaver Brook in the latter part of the seventeenth century. This was basically Puritan country at the time, a religion that forbade any celebrations, including Christmas and Easter, let alone Twelfth Night and Wren Day. All these holidays would certainly have been even more despised than they already were by the anti-festive Puritans if they had known that the festivals all had deep pagan roots and reached back to rituals that were performed in order to ensure the return of the sun in these dark hours. Safe to say, the rituals have worked. The sun has always come back.

• • •

I gave up on my wren hunt and skated on, passing along lanes and walls of tall brown cattails and skeletal branches of buttonbush, some of which were festooned with old fraying birds' nests. I could hear the unique sharp scrape of my skate blades rhythmically sounding out in the silence, and then soon enough, ahead of me I came to the large, brush-tangled beaver dam, and climbed over it.

I take it the beavers were even then at home and sound asleep when I reached their lodge. Beavers don't hibernate, but they do lay low for the season in their winter quarters. They'll collect edible branches and jam them into the bottom of their brook or pond just outside the underwater entrance to the lodge. Inside, they construct a stick platform, where they may feed on the bark of the branches they bring in. There is another platform higher up just in the dome of the lodge, and here the beaver family spends most of its winter days, all huddled together and sleeping the traditional long winter's sleep. They insulate their lodges with mud and are further insulated after heavy snows cover the lodge. The air gets thick in the sleeping quarters over the season, but they construct an air hole at the top of the lodge. And with all the insulation and their body heat, they can keep the temperature above freezing, even on the coldest of nights. Like bears, they fatten up for the coming shortage of food supply in the autumn, but unlike bears they store most of their energy reserves as fat in their large, flat tails. The Native Americans in these parts used to eat beaver tails, presumably for their fat content.

Along with a number of other mammals, the beaver has returned to this region after an absence of nearly 300 years. The local Indians used to use beaver fur for their winter wear, and ever since the eighteenth century their fur was used for top hats, which by the early nineteenth century were considered the height of fashion in Europe. The result was a vast increase in trade, first using Indian trappers and later adapted by the so-called mountain men, who would set out alone for the trapping season, and then once a year gather together to trade the skins and throw a big party known as the jamboree. After a few decades, beaver populations began to decline, first in the northeast, where they were once common, and then westward. They appeared to be en route to extinction until the fashion in hats declined and the trade ceased. In the 1930s, a pair of beavers was reintroduced to a wildlife sanctuary in western Massachusetts and slowly began to reestablish themselves. They eventually spread through the state and finally the entire Northeast. They returned to Beaver Brook in the mid-1960s.

Beavers were once an important aspect of local ecology and landscape. They backed up waters with their dams and flooded local low-lying forests. The trees died and wood-boring beetles, woodpeckers, and other hole-nesting birds such as wood ducks, took up residence in the trunks, and great blue herons established rookeries in the dead branches. In time the trees fell; the ponds dried up and, as a result, open meadows were created. Slowly over the years, what are known as early seral stage shrubs and herbaceous plants took hold, followed by sun-loving trees, and in time the native trees of the forests returned.

About ten thousand years ago, the native people of Siberia (mistakenly termed "Indians" by Columbus) began following the herds of grazing Pleistocene mammals such as the mammoth and the barren-ground caribou northward after the retreat of the glacier. They would have encountered the giant Pleistocene beaver at this time. And as is often the case, the giant beaver entered their folklore. The beast was known as Wishpoosh, and in contrast to our contemporary beaver was hardly a gentle soul. He would kill people who fished in the lake-sized ponds he created by his dams. Among western Native American cultures, however, our contemporary beaver was more benign. As with European traditions, he was associated with hard work and perseverance, and he was also a benevolent shape-shifter and could slay the monsters of the native mythologies.

• • •

Once over the beaver lodge I set out again, skating first in another wide loop that brought me close to the eastern bank, and then southward to a spot where the stream narrowed considerably. The water was running hard here and passed through two stone embankments, the remains of a ford that connected the farmlands of a family named Frost who in the nineteenth century held property on both sides of the brook. The ice was thin and clustered into fragments at this point, and the black waters piled through the channels in short but smooth wavelets. I had to climb onto the bank at this point and pick my way over the high ground and through the iced-over cattail marshes surrounding the narrows before I could get back to the thick ice.

In all likelihood, this was the spot where the family of Tom Doublet maintained their fish weir, and the same spot where Tom's father was killed. Beaver Brook in their time was open all the way to the Merrimack River so that migratory fish such as herring and shad could swim from the mid-Atlantic to inland ponds and the upper reaches of rills, swamps, and sinks, where the brook has its origins.

These so-called anadromous fish breed in fresh water but spend their adult lives at sea or in bays. Eels, which also used to swim through the weirs, would do the opposite. They are born in the ocean, in the far-off Sargasso Sea, and after the long voyage to the continents they swim upstream to spend their lives in freshwater ponds and lakes. After about eight years, they turn and swim back to the Sargasso Sea where they were born, to mate and lay their eggs.

The coastal people had access to this abundance of resources, and each spring when the fish were running, they would collect at certain spots along the course of rivers and streams to net the fish. But this wealth of a food supply also attracted the more violent and war-like tribes from inland areas. Spring was also a season of raids and war parties of the Iroquois and Mohawks, who were then known as the Magus, or Man-eaters. It was probably one of these raids in which Tom Doublet's father was killed. What actually transpired is unrecorded; the Maquas may have sprung upon him immediately, killed him with their war clubs, and then taken his catch. Or they may have parlayed a bit, perhaps argued, and then, by way of solution to whatever disagreement they may have had, murdered him.

A few hundred yards downstream from the ford, the brook widens and backs up at an embankment at a bridge. This was a bright winter day, but as often happens in this season, a northwestern wind developed later on that morning, and the skate to the bridge on the open reaches of the brook was not as relaxing as it had been in the more protected upstream areas. Nevertheless, I sped on, leaning into the winds and executing long, slow glides. And eventually in this manner, arms folded against the cold, I reached the bridge.

At this point I unpacked a pair of shoes, changed from my skates, and walked home on the road.

CHAPTER THREE

THE SNOW MOON

It was February and the woodchuck, were he with us now, could tell us what kind of winter it would be. But as it was, he had long since gone to ground and was still locked in his stone-cold sleep. Out in the fields to the west of the stream bank, the juncos were scattering like wind-blown leaves, and you could see the outline of the vole tunnels in the melting snows and the stories of the comings and goings of things in the night. On the fifth day of the month, an ominous red moon rose over the wooded ridges to the east. Great horned owls began their late winter caterwauling; foxes were barking above the brook, and the tracks of coyote packs stitched the trees together. It was a dead land waking; rain, ice under snow, rain, frost again, snow mist, and out in the world beyond the stream, the news was not good. Sunless days, freakish train wrecks, earthquakes, floods in the South, fires and mud slides on the West Coast, and all around the world, wars, and vast population shifts of refugees on the move, their cities bombed and local rebel armies rising, and everywhere the seas rising.

But then, in the midst of it all—Saint Valentine's Day—with all its hope of love and chocolate and its deep body of ritual and nature myth.

I went down to the brook in the early morning of the fourteenth. This was an oddly warm day, with mists rising over the melting snows. It had rained in the night, and all the storied tracks of the night wanderings of the local mammals had been obliterated, and the mist was so thick in some areas of the woods I could hardly see ten yards ahead of me. I could easily envision the flitting human figures there, appearing and disappearing in the foggy airs—old Tom Doublet, and Mary-Louise Dudley, who in 1722 was accused of witchcraft but died quite suddenly—and mysteriously—before she could go to trial.

This whole eastern section of Scratch Flat was said to be haunted by such figures. Quite apart from a mythical bear shaman who appeared from time to time in the hemlock grove above the brook, there were witches and ghosts in the land, and curses laid on certain tracts by Tom Doublet, who may have had shamanistic tendencies. Doublet was one of those historical figures who seemed to have the ability to end up in many stories of this region, including, in one account, a charge of sexual assault on a young woman on the Beaver Brook marshes. The case was unproven however, and eventually withdrawn. Doublet was not even on Scratch Flat at the time of the reported incident.

The rain had cleared a spot on the stream bank, but inasmuch as it was too wet to sit down, I leaned against the tree for a while and looked out over the marshes of canary grass, cattail, and buttonbush.

The stream was a silvery color, the ice was topped with about two inches of clear water from the recent rains, and some twenty yards downstream, a hole had formed in the ice. You could see the black waters of the stream below.

Saint Valentine, whose origins are more or less lost in the contemporary world of commercial valentines, was a third-century Roman Christian who could apparently perform miracles and thereby gained converts to the then new religion. His story was probably based on an earlier Etruscan personage with similar abilities and was later transferred to a Christian saint.

Valentine fell into disfavor with the Emperor Claudius II at one point and so, needless to say perhaps, was condemned to death. Regardless of whether any of this was true, and despite the fact that some branches of the Church did not think Valentine met all the criteria for saintliness, the legend endured.

The association with love and chocolate was a later addition to the holiday. The traditions of romantic love first appeared in the West in the lays of courtly love sung by the troubadours in twelfth-century France. Courtly love, which honored pure unconsummated relationships, was all the rage in the time of the twelfth-century Queen Eleanor of Aquitaine and her court and also in the society of her daughter, Marie de Champagne. The presence of romantic love was later a critical aspect of the romances of Chrétien de Troyes, and also, of course, King Arthur and his knights. But according to my source, Riggs, the first written association with love and Valentine's Day appears in Chaucer's *Parliament of Fowles* (1382) in which birds engage in a lively debate on the nature of love and attraction. At the end of the debate, Nature herself wins out and all the birds gather to choose their mates on Saint Valentine's Day. The poem ends with a song of praise for Saint Valentine.

The choice of the February 14 was no accident, however. As with all these festivals, the essence of the holiday reaches

far back into the pagan era. That is the day, according to tradition, and no doubt generally substantiated by the observa-
tions of country people of southern Europe, when migratory
birds return from Africa and begin their various mating rituals,
complete with birdsong and mating displays.

I had been hearing the early mating songs of cardinals,
chickadees, and titmice that same week, and also the periodic
howl of a seemingly lovelorn coyote who must have been separated somehow from his family pack.

A lonely howl of this sort, echoing through the halls of the
deep forest, would have haunted the nights of the early settlers
in this area. But among Native American cultures the wolf
and in the Southwest, the coyote, was a revered animal, often
connected with danger and strength and even considered part
human among some tribal groups. It was associated with the
creation and among certain Southwestern groups was known as
"God's dog." Among the Hopi, a character known as "Coyote"
is a trickster figure, able to shift forms, and he shows up in
many Hopi folktales.

But not so with wolves in Western societies. It's no wonder that the British settlers of the Beaver Brook lands would
have been terrified; they knew all the folktales. The earliest,
no doubt pre-Hellenic traditions, tell of whole companies of
wolf-like beings from Scythia, and there are numerous stories
of wolves and magical transformations in which humans are
changed into dangerous wolves. Greek mythology includes a
tradition that describes Zeus as a wolf-like beast-god. In one
version, the mortal Lycaon, seeking to test the omniscience of
the divinity Zeus, feeds him human flesh. As punishment for
tricking him, Zeus turns Lycaon into a wolf, the form he must
live in for the next nine years.

Among the Romans, the wolf had similar attributes, some
benevolent, as with the Capitoline wolf who nursed Romulus

and Remus, the founders of Rome. The Romans had a festival on February 15 known as Lupercallia that was associated with an earlier Greek festival in which the powerful wolf-god Zeus held sway. As with the Greek myth, the tradition involved human sacrifice in which anyone who consumed the flesh of the sacrificial victim would be turned into a wolf. After a period of nine years, the festival was repeated and those "wolves," who presumably were invited to attend the festival, once again consumed human flesh and were transformed back into human beings. It is unclear why the festival occurred in the middle of February, save that it is the time of year when migratory birds return and the hedgehog and other hibernating animals, including bears, begin to appear on the land after their long winter sleep.

The migration of birds among the ancients was a great mystery. No one was certain where or why birds disappeared in autumn and reappeared in spring. Some legends held that birds would fly off to the moon; others held that they dug into the earth like the bears and the hedgehogs. Aristotle had part of the answer. He calculated, accurately, from observation that cranes flew south to the Nile region, but he had it all wrong with the five species of swallows that occur in Greece. Having observed hibernation in mammals and reptiles, he calculated that swallows also hibernate.

With the advent of Christianity in Europe something had to be done with all these deeply rooted pagan traditions, and so the Church turned the wolf into a demon. And as a part of this transformation, the legends of the werewolf developed.

In what is now Russia, Sviatoslav Olgavich, the father of the epic hero Prince Igor, was a powerful ruler who was responsible for many victories for the Russian people in their constant defensive wars against the Scythians. He was a warrior by day, but at night he would turn into a wolf. Ethnographers cite him as part of the development of the werewolf myth, although the legend was no doubt enhanced by the belief that nearby Scythia was populated by a breed of wolfmen. There was also an early Russian mythic prince named Vseslav who would become a wolf at night. Animal transformations of this sort were common in ancient mythologies and were first recorded in the Mesopotamian *Epic of Gilgamesh*, when a goddess turned an innocent shepherd into a wolf.

In earlier, pre-Christian folklore, these wolfmen were miserable creatures who lived in their wolf persona but did not do any damage. But after the advent of Christianity, werewolves became hideous outsiders, hungry for human flesh. Significantly, early on in the journey through Hell, Dante and Virgil are followed by a she-wolf, a symbol, according to my friend Riggs, of the ancient pagan world and avarice and greed.

The wolf does not fare well in the Middle Ages, to say the least. The peasantry in particular despised the beast, who was, according to the church, in service to the Arch Demon, the Devil. This was not the best of times in Europe; there were waves of plagues, periodic famines, endless wars, and a repressive feudalistic social order, and all of it dominated and controlled by an all-powerful, inescapable religious institution. Wolves took much of the blame for anything that went wrong, which in some ways was understandable. They were seen feeding on the corpses of the dead during the plague years, for one thing. They killed cattle and sheep, and from time to time they may have killed people, and also spread rabies. And in Russia, according to tradition, they would give chase to families riding through dark forests in their troikas.

One can only imagine, given this ancient Western European association with wolves and their human/animal fellow travelers, the werewolves, the terror the English colonists must have felt on cold February nights when wolves began howling from the depths of the as yet uncleared forests. To the Indians, the call must have meant one thing, and was not necessarily frightening, but for the Pilgrims and Puritans who landed on the North American shores in the mid-seventeenth century, it meant quite another.

William Bradford's cold description of the wild world the Pilgrims encountered when they first landed on these shores sums it up:

"And as for the season it was winter," he wrote, "and they that know the winters of that country know them to be sharp and violent and subject to cruel and fierce storms . . . besides what could they see, but a hideous and desolate wilderness, full of wild beasts and wild men."

There is a debate among biologists about just who the current wolf-like residents of Scratch Flat and its surrounds are. The Eastern coyote is a much larger and more wolf-like creature than his Western counterpart, who, in comparison to wolves, is but a craven cur. According to some theories, the current "coyotes" of New England and Canada are related to the red wolf, which is an extant southern species and may have been the true wolf that the Puritans first encountered. They must have been very common, since the first Puritan settlers saw fit to construct a fence across the narrow spit of land that connected Shawmut, the peninsula that is now Boston, in order to keep the wolves away from their cattle.

Tom Doublet and his fellow Christian Indians who lived at Nashobah Plantation would certainly have heard wolves howling beyond the village clearings, and no doubt Tom's father or some local storyteller would have shared the old traditional native tales of the wolf with them. Bears, wildcats, owls, eagles, and other local denizens of the forest were the standard players in all these nature-related tales. Among these, Tom Doublet may have been told the story of Grandmother Woodchuck, who was the grandmother of Gluskabe, one of the pagan tricksters of the Eastern Woodland cultures. Grandmother Woodchuck was a wise old animal who doled out wisdom to her energetic and sometimes foolish grandson. At one point, for example, Gluskabe got in a tussle with a giant eagle who was the creator of all the winds of the earth. He managed to tie the eagle up in order to halt the ferocious winds that were plaguing the people. Grandmother Woodchuck pointed out that without the eagle wind, the earth would get too hot, so after some debate, Gluskabe untied him.

Given the fact that Grandmother Woodchuck is associated with the territory of the New England tribes to which Tom Doublet belonged, it follows that she may have poked her head up during warming trends to check the weather conditions. But as far as Groundhog Day is concerned, there is another origin. Among the Germanic tribes, February was the time of year when the local bears would emerge from hibernation.

Bear lore, of course, was as important among the central European tribes as it was all around the northern hemisphere. According to some folklorists, the German immigrants who settled in the American Midwest in the 1800s brought the bear tradition with them, but they transformed the bear into the more common woodchuck; thus, our tradition of Groundhog Day, which occurs on the second day of the month of February.

• • •

Rather than linger at my spot on the stream bank, I decided to tramp home the long way around, a loop that includes a quarter-mile walk along the wooded western banks of the stream and then turns westward through a wooded track. This was not easy going. The solid snow crust had weakened in the recent rainy warming trend and I kept breaking though periodically into the deeper base. At one point in the woods

I caught the scent of a skunk, or perhaps the musky smell of a fox marking his territory, but I could see no tracks in the rain-washed snows. Skunks do emerge at this time of year, along with possums, and raccoons. Larger mammals (except for bears) such as the bobcat, coyote and white-tailed deer remain active throughout the winter season. But the smaller mammals, such as the skunk, possum, and raccoon, tend to lay low and then emerge to feed when the weather is favorable. Mice are also out and about throughout the winter, and voles and shrews are active beneath the snow cover. In better conditions, that is, after a light snow, I can witness the amount of activity that takes place each night.

In time, bushwhacking through thickets, and halted periodically by the snow conditions, I finally came to a recognizable landmark—a huge boulder, or glacial erratic that once marked the northwest boundary of the Nashobah Plantation, the Christian Indian village where Tom Doublet lived until 1675, when the village was abandoned. Until developers came and cut it down, there used to be an immense white oak growing near this boulder, an ancient tree with low sweeping gnarled branches that curved downward toward the ground. This was all fields and pastureland back then, and on summer afternoons I used to climb onto the rock to admire the tree, and the surround of terraced fields that dropped down from the nearby Beaver Brook Road to the floodplain of the brook.

There are still oak trees along the brook, mostly red oak and black oak, but the great so-called wolf tree, which the local farmers left standing to provide shade for their cattle, is gone. Oaks, perhaps needless to say, are important players in the theater of legend and lore. In fact, they are without compare the preeminent tree in the whole body of European nature myth. It was oak wood that fed the sacred perpetual fires of Vesta in Rome; Jupiter and Juno wore crowns of oak branches; oak wood was used as the fuel for the holy Beltane fires in England, and was hallowed among the earlier Celtic cultures. The Druids, the Celtic priests, revered the tree; rituals were performed in the sacred oak groves; in fact, the word "druid" means "knowing the oak." Robin Hood and his merry men used to shelter under the vast Major Oak in Sherwood Forest, it is said. The tree, which is still standing, is believed to be about 800 to 1,000 years old. Not surprising, then, that the Christian Saint Brigid should be associated with the oak. She was an Irish nun who lived in the time of Saint Patrick and Columba, and was well known for her good deeds. She built a chapel on the site of a great oak that was associated with the pagan goddess also named Brigid, a healing spiritual figure who was connected to the sun and also apple trees and bees—a nature goddess in other words.

One of the ironies of the early Roman Catholic Church is the fact that the proselytizers who traveled through pagan Europe converting people inadvertently created a good record of the earlier pagan traditions. Saint Brigid, whose festival day is February First, is a case in point. The Christian missionaries traveling through the forests of Germany and Gaul would not attempt to stamp out the local religions by violence. They would simply point out that the local people were worshipping the wrong god and would substitute a Christian saint for the local spirit.

Closer to home, traveling in Gaul among the Parisii tribe in the region that is now Paris, Saint Denis came upon a sacred spring on the Ile de France that was worshiped by the Parisii. He explained that this was also a sacred place in his religion and the people should honor it in Christ's name, which they eventually did. The place of the spring is now the site of the cathedral of Notre Dame.

Saint Boniface, who converted the Germanic tribes, came up with a more radical means of convincing the pagans to join his religion. In order to complete his work, he personally cut down the famous Donar Oak in Hessen that was venerated by the local tribes.

Such symbolic gestures take place, even in our time. One never knows when one is living in the midst of vast historical and ecological changes, and yet viewed from the longer perspective of geologic time, we are now living in such an era. There are still oak trees growing in the strip of land along the west side of the brook, protected currently by local laws controlling the uses of land within the watershed of Beaver Brook. But years ago, when the whole tract of land between Forge Village Road and Beaver Brook was all fields and pastures, I used to survey the surrounds from my perch on the boulder and consider the changes that have taken place on the area known as Scratch Flat over the last 15,000 years. I wonder sometimes if, in a metaphorical sense, the slaughter of the great white oak by the boulder may have marked the coming of another climatic change to this patch of the North American continent.

After the retreat of the glacier, this land was forested with oak and maple, hickory and ash, as well as a few species such as the now extirpated chestnut tree. That forest endured for 8,000 years, then in the mid-sixteenth-century there came a force that was almost as powerful as the glacier as far as changes in the land were concerned—the arrival of the Europeans, a group who, unlike the former native residents who had occupied the land for some 10,000 years, tended to remake the land to suit their own needs, rather than shape their culture according to the ecological makeup of the place. The English began to settle the lands around the brook in the 1650s and started clearing the native forest as soon as they arrived—a prodigious amount of work when you consider the size of the trees, and the vast numbers of rocks, some of them too big to move—that were buried in the soils by the late great ice sheet.

The forest clearing that began in the mid-sixteenth [1744] century was thoroughly completed by 1850. From the ridge above the brook in those years, you could look southeast and almost make out the spires of the churches of Concord and the villages of the metaphorical Thoreau Country, or look west to the seminal heights of Mount Wachusett, which so inspired Henry Thoreau in his wanderings. But starting in the early 1800s with the opening up of the wheat growing lands of New York State, the topsoil of their land having worn out, the New England farmers began to move westward. The trickle of desertions became a river (or, literally, a canal), with the completion of the Erie Canal in 1825. With the retreat of the farmers, and the abandonment of the open fields, the old forested landscape, which had been lurking for centuries in the seed bank of soils, began to sprout. First appeared the sun-loving plants such as dogwood and grey birch, along with a thick understory of shrubs, and then the old oaks and the hickory and also the white pine, which in its original form was what was known as an emergent species. They grew singly throughout the forest, but were tall and straight and reached above the canopy of the deciduous trees. These sun-seeking pines grew quickly and outstripped and shaded the struggling oaks and hickories, and soon enough, that is, after a hundred years or so, they became, at least in this region, the dominant tree species, although the oaks and the hickories and maples also managed to struggle back.

In Thoreau's time, eighty percent of the whole northeast was open land, but by the turn of the twenty-first century the reverse was true; it was eighty percent forestland. But that is not to say that open space in our time is not disappearing at alarming rates, and the slaughter of the ancient white oak, at least in my mind, could stand for it all.

• • •

There is now no way I could clamber up onto that boulder to view the landscape. The almost vertical walls of the rock are surrounded by a dense, thorny, and impenetrable fence of multi-flora rose and euonymus, an alien plant introduced into this country as an ornamental, which now has escaped and multiplied, and is crowding out the native understory shrubs such as blueberry. One wonders what nineteenth-century naturalists such as Henry Thoreau would think of such conditions.

Actually, we do know in a sense what Henry would think. He was a lover of wildness, and to him the cultivated fields were altered landscapes, not half so intriguing as the remaining woodlots. He would have, no doubt, railed against the invaders and longed for the ancient wilderness of oaks and maple— the "desolate wilderness, full of wild beasts and wild men," as William Bradford wrote. Thoreau said as much in one of his most celebrated epigrams: "In wildness is the preservation of the world."

Henry Thoreau probably did wander up along this ridge above Scratch Flat at some point in his rambles. Ralph Waldo Emerson had a brother, Bulkeley, who was in some way mentally challenged, although no one recorded or even knew, perhaps, what was wrong with him. He would shout "inappropriate" things at the dinner table according to one account, which suggests that he may have suffered from Tourette's syndrome or autism, or some such disorder. Whatever it was, the Emersons farmed Bulkeley out to another household, and for a while at least he may have worked or even lived on the local town poor farm that was located on the western banks of the brook not far upstream from my lookout.

Henry seems to have had an open mind about outsiders of this sort. Periodically, he came up to the Beaver Brook area to visit with Bulkeley, which in his day may have been a pleasant

ramble since the dense thickets of woods were all cleared. I
have done this particular ramble myself, albeit in reverse, and
it took me all day, burrowing through thickets and crossing
swamps and marshes and running streams. The last time I
did it, I attempted to retrace Henry's most direct route to the
local poor farm.

This was in autumn, a sultry warm day after a heavy night
rain, and the earth was redolent of moist earth and leaves,
and every once in a while the musky scent of what the locals
around here used to call shittum bush was in the air. I had
started early, but with all the traipsing around the backyards
of new houses so recently built in the returned forests of the
American past, and forging marshes and streams, I arrived at
the outskirts of the Estabrook Woods in Concord only late
in the day.

There is an abandoned road running along the western side
of the Estabrook Woods, and, cutting across a few backyards
and disregarding formidable "No Trespassing" signs, I event-
fully got to the road.

This short unmaintained track has a deep history. There was
a lime kiln along the road at one point, and you can still see the
foundations of the house and barns of the former Estabrook
family, who, along with so many early nineteenth-century
farmers, deserted their land in the 1820s.

I was tired by this time, and this was easy walking for once,
no rocky outcroppings, marshes, or thickets, and I strolled
along slowly, daydreaming about the place as it was, not as it is.

In the midst of one of these meditations, I saw approaching me from the south the figure of a gypsy queen. Unlike so
many women in our time, she was wearing a full skirt, a light
beige blouse, and a vest, and she was walking along with an
ambling dreamy pace. I was able to recognize her even before
I saw her face. She was a local craftswoman and artist from the

town who was, as are so many in these parts, an avid follower of Saint Henry Thoreau, the wild spirit of nature.

Apparently, she recognized me at about the same time; she stopped in her tracks, stamped her right foot, and spread her arms in welcome. This was Kata Grant, an old friend of mine from the days when I lived in Concord Center.

Kata was a fluent, rambling, effusive, and unending talker. Set her off on a subject that she is familiar with or even associated with, and she will travel on for half a dinner, nonstop. But unlike not a few types of this sort, what she has to say is usually intriguing.

She had come to Concord from the Mill Valley in California, via the Berkshires in Massachusetts, where she had married, then divorced, and then fled the country—for Morocco, I believe. She then came home and lived among the Hopi for two years and then finally moved to Concord. It was a case, she told me once, of *Tuwanasaapi*, a Hopi word meaning "place where you belong." Although she had made a living for a while as a graphic artist, she was best known as a basket maker and actually had some of her traditional baskets purchased by museums, including the Smithsonian. By the time of the quincentennial, in 1992, she was sent to Dominica to learn how to make the traditional square baskets used by the Awaraks, the people who Columbus first encountered. Kata also had a refined knowledge of local plants, as did Henry Thoreau. It was she who introduced me to the habit of checking the dates of my own nature jaunts against the same dates in Thoreau's journals. This turned out to be an excellent way not only of connecting myself to Mr. Thoreau, but also of tracking the changes in the land that have taken place since his time, especially those in the plant community. In recent years, the Thoreau journals have been used to track global climate change by comparing the blooming times in Henry's day against those in our time.

The flowering plants on Scratch Flat, which is about fifteen miles northwest of Walden Pond, have advanced by two weeks or so, I have noticed.

You would not know any of this in February. Unless you know how to read animal tracks and tree buds, the world here is transformed into a cold, desert-like, whiteout landscape.

• • •

The day after Valentine's Day there was yet another snow-storm. Not a particularly heavy snow and later in the day I went back down to the brook using the path through the sportsmen's club, since I calculated I would be safe on such a day—no one would be around save perhaps for Riggs, who also sometimes goes for walks in the middle of snowstorms.

The tree boles were black and the lower dark limbs of the pines reached out toward me, claw-like. Were I a believer in such things, I would say that there were evil, witch-like wood spirits about in that dull late afternoon gloom. This was exacerbated by a surround of hideous groans and eerie whines of the wind, creaking the branches overhead. At one point, I was stopped in my tracks by a shuddering thunder-like crashing. I glanced up just in time to see the gray-brown blur of two or three deer fleeing. Other than that, I was alone, not a bird, no squirrels, no dire wolves, or woodland caribou, saber-toothed tigers, or woodland bison such as might have been seen here 10,000 years ago and whose ghosts I like to imagine still haunt this land.

As it turns out, in 1860 there was also a healthy snowstorm on February 16 in Concord, and so, of course, Henry went out for a walk. He was, as he wrote somewhere, the official inspector of snowstorms. His description in the journal entry for this date is highly ornamented with visual detail, the dark ink etching of black branches against the white pages of the snow, and the obscuring blanket of the snowy fields and streams. It was a far more benevolent environment than the one I had been wandering through; he crossed woodlots and fields and pastures and makes no mention of ominous groans and screeches and nothing of the ghosts of ancient extinct mammals. Mystic that he was, Henry was also an acute scientific observer of the natural world. Landscape for him was another matter. He tended to set his imagination free on his local expeditions and fancied Mount Olympus and Ararat in the modest hills west of Concord, and the immortal gods of ancient Greece atop Mount Katahdin in Maine.

The snow that began that day continued through the night and ended in the early morning, followed by a warm sun. It was a damp but light snow that clung to all the branches and hedges of my garden, and it had turned the world into one of those mythical more benign, well-lit faerie tale scenes, with the stone walls and the pointy firs capped with Chinese helmets, and drooping white arms of the hemlocks. And so, following the same path I had used the day before, I threaded my way down to the stream bank and headed to my place beside the brook. I was hoping to find some interesting animal tracks, but except for the bounding signs of the two deer that had fled the day before, there was not much to see. Either all the animals had lain low in the storm or the early dawn snow had covered them.

I used to know a couple of outdoorsmen who were expert trackers; in fact, they both taught classes on animal tracking in different parts of the state. I used to go out with one of them from time to time to see what I could learn. He, of course, knew which animal left which tracks and the meaning of the distance between the footprints, and the depths of the prints, and which tracks indicated halts, or bounding leaps, or a walking pace, so that you could more or less read what the animal was doing in the course of its day. He could also identify other physical signs—the fact that rabbits nip off shrub twigs cleanly, whereas deer will chew them off, leaving a ragged end. He also taught me a lot about bear signs, such as the fact that male bears will reach up a tree trunk and leave claw marks—the higher the scratch, the bigger the bear, a sign that other bears were able to read.

Hunting cultures are expert trackers, and no doubt Tom Doublet would have known all the signs that various species of mammals left behind, not to mention the contents of droppings, which is to say the presence of berries or seeds or feathers and fur, for example, and also the smell of musk left around by foxes and also skunks and other members of the weasel family.

In many cultures, tracks, especially human footprints, held a sort of contagious magic, so that by manipulating the track you could exert control over the person or animal who left the print. Certain tribes in South Australia believed that they could lame a person by driving a sharpened bone or a piece of quartz into the footprint. The same method was used in Germany even up to the mid-nineteenth century. People would drive a nail into the footprint of an enemy, and in parts of France and England it was believed that if you stuck a knife into the print of a witch, she could not walk until the knife was withdrawn. On a more benevolent note, in one Slavic tradition, would-be brides would collect the soil of the footprint of a potential husband, pot it up, and plant flowers in the soil.

I've never run across information on traditions of this sort of magical transference among the Native Americans, but in certain hunter tribes the world around, in order to improve or assure the success of the hunt, dancers would reenact the chase, imitating the trail of the animal in dance steps, finding it, and spearing it. This sort of sympathetic magic has ancient Paleolithic roots. One theory of the cave paintings of the Dordogne and surrounds is that the act of painting, or the painting itself, would locate the prey and assure the success of the hunt. That theory is currently debated, however, and may be out of fashion in our time; the actual purpose is still not known. One argument holds that there was no purpose; the paintings were simply a Paleolithic version of art for art's sake. They painted their world, which, of course, was dominated by horses, maned lions, bison, and woolly mammoths.

This particular snowstorm had a softer, moister scent than early snows, a definitive springlike feel. Over in the hemlock grove behind the house a few weeks earlier, I had heard the ghostly hooting of a great horned owl. They mate in February and lay their eggs late in the month, and the chicks hatch in March.

In the woods that morning I had heard the twittering of a purple finch and the little nasal honk of a white-breasted nuthatch. Buds were swelling and a pool of black water had formed near the bridge at the north end of the stream.

With my work as inspector of snowstorms completed, I turned west and hiked back up to my house through the fresh unmarked snow. At the head of the driveway leading to my house, a flight of birds burst suddenly out of a circular planting of hemlocks and whirred overhead. I spun around just in time to see a flock of blackbirds rolling across the road toward the stream. I couldn't get a good look, but they had all the marks of red-winged black birds. It was just one more clear sign that winter was beginning its slow, but determined, march toward spring.

CHAPTER FOUR
THE WORM MOON

March came in with a chill wind that skidded over the icebound stream, a sharp, clear, biting wind that pierced layers of outerwear and froze the bones. And yet, by the beginning of the second week, the wind dropped and I was able strip off my gloves and scarf and open my coat to the warm sun and settle on the dry grasses of the bank, my back against the trunk of a red oak. Black pools of swirling water had opened in the ice, and along the shores on bare rocks, I found little congregations of stoneflies, the semi-aquatic insects that sometimes emerge to bask on rocks even in mid-January. A little higher up on the banks at the base of another oak tree, I saw a scattering of pepper on the snow that seemed to have the ability to jump or pop up. These were the tiny snow fleas or springtails, primordial insect-like species in the genus *Collembola*, which, cold-blooded though they are, have the ability to become animated even on chilly late February days.

On the swampy downstream marshes, the skunk cabbage spears were well advanced and the false hellebore was unwinding. Pussy willows swayed in the light airs, and having by this time arrived in full force, the advanced guard of the red-winged blackbirds was sounding out all across the marsh and were foraging near the spot where the skunk cabbage grows. Purple finches began to sing early in the month; the starlings and the cardinals and tufted titmice, the chickadees and the Carolina wrens, had long since joined the chorus, and at dusk, up in a meadow near the road I heard the nasal, peenting call of a woodcock, followed by its warbling whistle-like song as it descended to its ritual dancing ground only to start all over again with its nasal calls. Willow branches were just beginning to show shades of yellow, and the stems of osier dogwoods were bright red and looking for all the world like Spanish dancers in the snow. Muskrats had emerged and were feeding along the shores, and wood frogs were beginning to collect in the upland vernal pools.

It was all chill air and scented earth and swelling buds, birdsong, and bright winds, and the hope of things to come.

Later in the month I went out on one of those bleak March mornings and headed down through the woods, taking the wilder untracked route that I would sometimes follow. There had been another light snow, and it seemed a good day for animal tracks. A brace of coyotes had crossed the little meadow where the woodcocks would gather each evening and then had threaded their way through the trees, going who knows where, but not stopping to investigate anything; they were like those dogs that you see sometimes in rural areas trotting on in a determined way on a single-minded mission—to home base, perhaps. Later, I saw the tracks of a small herd of three or four deer—a more disordered traveling, and then, closer to the brook, I saw the long-padded footprint of a fisher, a large weasel that has returned to these parts after an exile of some 150 years or so.

Earlier that winter in this same area, I had followed a bobcat for about half a mile before giving up, and on another good tracking day I had seen a large, rounded animal footprint that wound off through the woods in a single line. I had seen tracks of this sort in the American West and also in Belize and Costa Rica. It was clearly a large cat, in this case a mountain lion, an unlikely event but not impossible. Reports of mountain lions here, not thirty-five miles from Boston and points west, had been coming in to the state wildlife offices for years and had been summarily dismissed—legitimately, in most cases. But serious naturalists had also reported undeniable signs of the presence of cougars, and two people I know, including my own daughter, who has seen cougars in the West, have spotted them. I lost this particular track in a tangle of briers, but no matter; I'm not sure I would have wanted to come upon the beast in any case.

The most interesting trail I came across that day was the single file line of a curious fox. It had found something worth digging for at the base of a tree. It trotted on for a while and then stopped to sniff some unidentifiable (to me at least) thing in the snow. It left its mark by peeing at the base of a rock, trotted on, stopped, turned right, and galloped through a clearing. Checking for other tracks I could find no reason for this detour, and in any case Sir Fox soon resumed his course.

I was curious as to whether this was perhaps the same large fox that had gotten into the habit of helping herself to my neighbor's chickens from time to time. My sunny breakfast alcove overlooks the garden and we would periodically see the fox crossing the yard, a dead chicken held high in her jaws. She was probably a vixen, carrying dinner home to her kits.

Eventually the fox I was tracking ended up in the same place I was headed—Tom Doublet's landing. I was interested to note that before moving on, it had sat for a while in apparent contemplation of the view. One would not put such things past a fox; they are known to be intelligent, and who knows whether they also have a refined aesthetic appreciation. Foxes the world around are universal symbolic animals; fox tales range from Africa to Europe, and east to Japan. In almost all these tales they are imbued with the same character, that of sly and wily tricksters.

My understanding from town histories is that there was a slave named Mingo living at one of the farms on the west side of the brook in the 1750s. He was a native of West Africa, and although there are no records of his transport that I could find, before he arrived at the local farm he had probably been living in the area of Boston known as New Guinea on the backside of Beacon Hill.

The enslaved people of the Puritans enjoyed more privileges than the plantation-based slaves of the South. Puritans ranked them as "servants," and so the slaves they owned were more integrated into the family and often ate at the same dinner table. They worked side by side with the local white farmers. The family that owned Mingo, the Caswells, had two young children, Adam and Eve, who—I'm guessing here—likely spent a lot of time with Mingo, who I gather first came to the family when he was a teenager. Mingo would have brought with him a host of local West African folktales, which he probably shared with the Caswell children on idle winter nights.

One of them may have been the story of a fox, originally based on the local West African pale fox, which was transferred in the slave culture of America to the red fox, the species I was tracking that day.

In the now controversial Uncle Remus tales collected by Joel Chandler Harris, it is the rabbit who is the trickster who outsmarts Brer Fox. Brer Rabbit is probably based on another West African animal known as the bush baby, and he is hardly the most honorable character. He's violent; he's a hustler and a liar; he's cruel and he's lazy. In one story he manages to get a hold of Brer Fox's head, and he wraps it up and gives it to Mrs. Fox as a gift.

The stories, as transposed by Harris, are based on the original African folktales. Chandler heard them from slaves on his own plantation in Louisiana, but other folklorists in the period recorded more or less the same stories. Up until the late 1950s, the tales were much loved by children of all races. I grew up with them and still have an old first edition copy of *Uncle Remus*. But by the time of the Civil Rights movement, the racist elements in the stories were criticized. The tales themselves were not necessarily racist; in fact, Harris was associated with a group of folklorists who were collecting traditional African tales in the New Orleans area in order to preserve the African heritage.

The criticisms were based on the fictional character Uncle Remus, who narrates the stories to a little white boy, and were written in the southern African American dialect of Harris's time. The African American writer Julius Lester, addressing the racism in *Uncle Remus*, solved the problem, or tried to, by taking Uncle Remus out of the story and just relating the original folktales.

The reversal of roles—the rabbit as a wily, sly character and the fox as dull-witted—is a unique twist. But there was a rich body of fox lore among the Native American cultures in the Northeast in which the fox was benign, wise, and often helpful. And among Western tribes, foxes were believed to be associated with the creator, an assistant of sorts, and among some groups he plays the traditional role of the trickster–along with his fellow canine, the coyote—whose tracks I had also seen that morning. These same attributes occur throughout European cultures as well, although in some tales the roles are reversed, as they are in some of the Brer Fox African tales.

One of these reversals occurs in the Nun's Priest's Tale in Chaucer's *Canterbury Tales*. In the story, the vain, colorful cock, Chanticleer, has a dream in which he is attacked by a strange monster with a bushy tail and a pointed nose. He wakes his favorite consort, Lady Percolet, and they engage in a long dialectical sophist debate as to whether dreams are real or not, each citing the classic philosophers of antiquity. Chanticleer eventually goes back to sleep and by morning has forgotten his dream.

But out in the chicken yard that very morning, he meets a handsome animal with a long nose and a bushy tail whose name is Reynard. The two get into a pleasant conversation, during which Reynard praises the beauty of the bright feathers of Chanticleer, who is flattered and honored and enchanted by the praise. But then, of course, at an opportune moment Reynard grabs Chanticleer and makes for the woods. The alarm goes up on the farm, and everyone—and every animal—gives chase, the farm wife and her daughters, the pigs, the cow, the dog, the cat, and even the bees. Seeing this, Chanticleer gets his chance. He praises the speed of the fox and tells him he can easily outrun these fumbling pursuers, and he suggests that Reynard call back and mock them for their inabilities, which the vain Reynard does. As soon as Reynard opens his mouth to call back at his pursuers, Chanticleer flies into the trees.

This story actually occurs in earlier Medieval French tales and probably can be traced back to the legends of bears and wolves that were told around cave fires in the Paleolithic Era. From these long forgotten, obscure tales, the fox made its way up into the Christian era in which, according to some versions of the stories, Monsieur Reynard, once the smooth-talking trickster, was personified as an agent of the Devil, along with goats and black cats and snakes.

This darker aspect of the fox personality may predate Christianity. After the invasion of Gaul, Roman chroniclers, including Julius Caesar, report the barbaric custom of the giant wicker basket in the shape of a man into which all manner of animals, including foxes, and also, according to the Romans, criminals, were packed and then burned in a bonfire. This was all part of a Druid religious festival, later associated with the midsummer fires of the solstice. There was also a Celtic legend that held that witches could turn themselves into foxes, which may be part of the explanation as to why this otherwise popular but tricky animal character had to be sacrificed.

When he was in his early twenties, my father lived in China for three years, teaching English in Shanghai. This was long before the Revolution and he became interested in the wealth of Chinese magical thinking, and the myths and legends of the local people of the countryside in the pre-Mao era. While he was there, he traveled to Japan on his summer holidays, where he grew interested in the work of Lafcadio Hearn, the Irish-born writer and wanderer who in middle age ended up taking up residence in Japan and became an authority on Japanese folktales.

After he came back from China, my father went to graduate school and ended up doing his thesis on mysticism in the writings of Lafcadio Hearn. He had a huge collection of first edition books by and about Lafcadio that I inherited after he died and from which I learned a lot of Japanese folklore.

Hearn was born in Greece of a Greek mother and an Irish military man. He moved with his father back to Ireland, where he was raised by strict Catholic aunties. When he came of age, he fled the country and lived for a while in the United States, mainly in New Orleans. He loved exotic cultures, spent time in the West Indies, and then sailed to Japan. When he first saw the country from shipboard, even before the vessel landed, he announced to a friend that he had found a home and intended to die there. He eventually married, changed his name to Koisumi, and became the interpreter of Japanese culture for the West.

When I was growing up my father used to entertain me with occasionally gruesome bedtime stories that were collected by Lafcadio. One of these tales—actually a series of them—dealt with foxes, specifically with the Japanese tales of Kitsune, a goblin-fox.

Kitsune is, as with most foxes, a highly intelligent spirit who grows wiser with age. He or she is often helpful, but can also be vengeful and filled with the usual tricks and disguises. She is also a shape-shifter and can appear to mortals as a beautiful woman.

One of the earliest of these fox transformations occurs in Japanese traditions around 500 BC. The story tells of a man who dreamed for years of finding a stunningly beautiful woman for his wife. Walking one day on the marshes, he met such a woman; they fell in love and married and had a son. But on the same night as the boy's birth, the husband's dog gave birth to a puppy who, as he grew older, became hostile to the lovely bride. She begged her husband to kill the dog, but he refused. One day the dog attacked her outright, and the beautiful lady transformed herself into a fox, leapt over a fence, and fled. Seeing her in flight, the husband called out, begging her to return even though she was now a fox. She was the mother of his son, he shouted, and should come back. He told her she would always be welcome. So every night thereafter the fox returned and slept with him as a woman and turned back into a fox at daybreak.

Some of the foxes in Japanese folklore are white and are especially good at bringing down the greedy or cruel Samurai by their tricks. They were popular characters in Kabuki theater and art, and fox goblins have even made it into our time in the form of video games and manga cartoons.

I thought I saw one of these fox goblins along Beaver Brook one autumn evening. I was headed down the cart track to the landing at dusk, and just before I stepped out of the wall of trees into the clearing, I saw a dark form sprint across the landing. It had the shape of a running fox, low and speedy, but it was black, like a shadow, and when it cleared the open space it disappeared into a thicket of brush and tree trunks on the south side of the clearing. It may have been a real fox, but it seemed to me that it fled through the dense thicket and the picket fence of tree trunks without disturbance, a little like a wisp of dark smoke. I stood wondering if I had actually seen a fox at all, it happened so fast.

. . .

Spring Equinox that same year was a hideous day. There was not so much as a hint of light in the morning sky, a leaden shadowy pall stretched over the open fields, and the forest was a gloomy fog-haunted hall, grey on grey, interspersed with the black pillars of the tree boles. A frozen mist fell, and there was a dankness in the air, the sort of northern European atmospheric condition in which it is impossible to get warm—except in front of an open fire. I went out anyway, well bundled but shivering nonetheless and made my way through the slushy forest floor and stood leaning against an old oak on the north side of the bank. The stream was open—a black ribbon winding through the brown walls of the winter-killed cattails and button bush, and yet, in the midst of this grey/brown stage the male red-winged blackbirds were calling. Their song was somehow warming—a reminder that this was, in spite of it all, the first day of spring.

Almost any day in later winter from my post I would see the beautiful black and white form of a hooded merganser floating by, the bright white of its hood standing out in sharp contrast to the dark waters and the lion brown hue of the marsh grasses and the cattail stems. I also saw otters swimming by from time to time. On one of those days, one of the otters spotted me and began chattering. I made kissing, squeaking sounds back, and it turned and floated on its back downstream with the current, squeaking and chattering.

For a short period of time in those years, I had actually gained official permission to invade the off-limits territory of the local sportsmen's club because I wanted to make a study of the otter population that seemed to have established itself along the brook. I had even offered to join the club but was informed by the president that that would not be necessary; I could go down to the landing as long as there were no cars in the parking lot.

My otter study never quite panned out since they were so hard to locate. I had seen their tracks and the mud slides on certain sections of the brook, and for a while one or two of them had taken to using my sacred sanctuary as an outhouse. They left scat around the edges of the clearing, complete with fish scales and fish bones.

I had also seen a strange running groove in the snow-covered ice one winter running all along the stream. I was curious to know what could have made such a trail; it looked like a single ski track, but save for a few brush marks in the snow, there were no signs of ski poles. It might have been perhaps the trail of the dreaded Apch'inic, a snake-like monster of Native American lore that lives underwater, usually in lakes or ponds. But I had heard no accounts of Apch'inic in recent years from my local Native American sources. One old man told me they all went away when Jesus came. This same man also told me that otters were supposed to be the "winter bringers" among his ancestors. But they too—that is his ancestors—had gone away, taking all their stories and the "real" otters with them. (He did not say what kind of otters I was seeing; they seemed real enough to me.)

I called a tracking friend of mine who told me the trail was in fact made by otters. Whether for entertainment, or just to get wherever they were going in a hurry, he said, otters would "paddle" along the ice with their paws, creating the groove that so puzzled me.

• • •

The vernal equinox, March 21, is the date when the night and the day are of equal length. Technically this occurs when the sun crosses the ecliptic, an imaginary perimeter in the sky that marks the apparent path of the sun over the course of the year. The sun crosses the ecliptic twice a year, once in fall and then again in spring. For the ancients, who were inveterate sky watchers, the summer and winter solstices must have been fairly evident. But to determine the two days in the year in which the night and the day were of equal length must have taken some sophisticated measurements. It probably was dis-covered by the earliest astronomers in Sumeria.

But with or without accurate astronomical measurements there were other earthly signs: the greening of plants, the return of singing birds, and the voices of frogs.

One of the more interesting and dramatic evidences of this ancient vernal festival occurs among the Mayan cultures, who were more accurate astronomers than many of the sky watchers of Western cultures. The great pyramid temple at Chichen Itzá, in Quintana Roo, Mexico, is situated in such a way that on the day of the vernal equinox the image of a giant snake, the feathered serpent Kukulan, which is created by the shadow of the descending late afternoon sun, slithers down the stepped sides of the temple and illuminates the head of a sculpted fanged snake at the base of the pyramid.

Snakes, which often reappear from hibernation to bask in the sun around the time of the equinox, also appear in worldwide spring celebrations. In Celtic and northern European cultures, the snake is associated with the fertile Earth goddess, Ostara, along with many of the traditional spring symbolic creatures of our time, including the hare and the cosmic egg. Two of these items, the egg and the serpent, can be found in another Native American monument, the great Serpent Mound of the Miami or Adena people in what is now Ohio. The mound is generally believed to have been a symbolic solstice monument. The 1,500-foot curving mound takes the form of a serpent, who faces west and is oriented toward the sunset on the day of the solstice. The mound is sculpted into the form of a snake swallowing an egg—presumably swallowing the sun. The structure may be based on the locally common black snake, which often climbs trees to rob eggs from birds' nests.

In Western cultures the spring equinox was associated with a long line of important gods and demigods. In the basic outline of related cultural tales, a young man is either killed, having been first tied or hung from a tree, or sometimes dies by accident. After his death, his mourners, or grieving goddesses, bury him in the ground. But he comes up again after three days.

The archetype of this theme appears in Sumeria in the form of a shepherd god named Tammuz. He was beloved by a powerful mother goddess Innana, as well as his sister. At one point in the earliest narratives, Tammuz is hung from a tree and tortured to death by demons and sent down to the Under-world. Innana, who was herself doomed to the Underworld for a while, rescues him and he comes up every year in spring with the returning vegetation.

Osiris, the Egyptian god of the Underworld, endures a similar fate. His brother Set, the god of war, killed him and in one version cut him up, and in another sealed him in a coffin and sent him down the Nile. His sister/wife, Isis, found him, reassembled him, and buried him. But he came up again, as he has every year since, in the form of the renewed vegetation of spring. Even while he was alive, he was reported to have had green skin, an indication, one supposes, of his close relation-ship with plants.

Later in history, in Phrygia, a young mortal man named Attis suffered a similar fate. He was a beautiful young lad, beloved by Cybele, the powerful Great Mother. Attis was the child of gods who abandoned him shortly after his birth to be raised by a goat. He was later sent off to a city in Persia where he was to be married. Cybele appeared at the wedding in all her splendid glory; Attis fell into shock at the sight of her, went mad, and castrated himself. In another version, he was out hunting one day when he was attacked by a wild boar and emasculated. After his death, he was buried, and then, after three days, rose again from the dead. His resurrection is honored in the form of an anemone, or a red crocus, which blossoms around the time of the spring equinox.

The Greek demi-god Adonis suffered the same fate. He too was gored by a boar and after his death rose from the dead, along with the rebirth of vegetation. After his death, an Adonis cult developed among Greek women in which short-lived plants were potted and left to die in the hot sun. The women would then mourn his symbolic death in a public display of grief.

Finally, 2,000 years ago, the god Jesus was hung from a tree, was buried, and, three days later, rose again on the spring equinox. But unlike his predecessors, Jesus was an historical figure, and the question arises as to how he came to so closely resemble these mythic precursors, as well as the Roman deity of Persian origins named Mithra, who was born on December 25 and died in the spring.

The Mithraic cult was popular in Rome in the time of Christ, and the myths and legends of the earlier pagan gods, not to mention a host of Germanic and pagan Roman folklore associated with vernal rebirth, would have been well known throughout the Middle East and Rome. The great Christian proselytizer Saint Paul, who traveled all through the Middle East and Greece spreading the Word, may have purposely transferred these myths to the tale of Jesus wherever he met resistance from pagan communities, which he apparently did in some cases. After a sermon at Ephesus, he was reportedly shouted down by a crowd, chanting "Artemis" repeatedly.

The other possibility is that the early Christians, who were drawn from the lower, generally uneducated, classes, would themselves have placed the miracle of the rebirth of Jesus in the camp of these earlier traditions.

Recent Christian scholars have cast doubt on this theory of the hanged god and his death and resurrection, especially when it comes to the Adonis cult, which was in vogue at the time of the birth of Christ. But this does not diminish the historical fact of worldwide myths celebrating the return of vegetation and life after the death of the winter season. Easter itself is drawn from the Celtic holiday Ostara, which was named after a Germanic earth goddess. And there were many animals and symbolic plants such as the lily and the dove that became connected with spring traditions, including the previously mentioned serpent, which along with singing birds and frogs is a logical conclusion. Plants die in autumn, birds disappear, snakes hibernate underground, and on the first warm sunny days of spring snakes reemerge along with the birds and the mammals and the green plants.

• • •

Later in the spring, picking my way along the banks of Beaver Brook or canoeing past high hummocks, I would often see large, black water snakes sunning themselves. There were no snakes on that cold spring day, however. Nor any flowers, save for the budding early flowers of skunk cabbages and the green leaves of hellebore, two of the earliest of the greening plants of spring, along with the flowers of hepatica that bloom sometimes even before the snow melts. I had seen a small stand of them half buried but showing through a surround of snow like a swathe of paisley a few days earlier behind the back wall of my garden.

Late in the day of the equinox I had left the bank and was walking up through the shadowed woods when a shattering caterwauling followed by a familiar hooting broke the silence. Anyone unfamiliar with the natural history of this land would assume that a claw-winged demon was about to descend and grab me, but I recognized the noise right away. It was the call of a barred owl. I had heard them earlier that week, booming out of the forest across the road from my house. The season of owls was upon us.

As early as late January the great horned owls began to nest in the hemlock grove to the northwest of the garden, and then in mid-March I heard the descending whinny of a screech owl, followed a week or so later by the calls of the barred owls, which nest in the swamps on the western shores of the brook.

Of all the birds that inhabit the fields and forests of the world, owls have probably more legends associated with them than any other avian—and not always pleasant legends at that.

I had read an account in Axel Munthe's book *The Story of San Michel* of an ominous event that took place in a cemetery in Rome in 1910. One rainy night Munthe, who was a well-known and popular doctor in Rome, was involved in a somewhat nefarious transfer of bodies from a grave in the cemetery at Porta San Paolo. He and the gravedigger were hard at work when, out of the gloom, from behind the Cestius Pyramid, a big owl began to sound off. The grave digger crossed himself immediately. Munthe was a great lover of owls and birds in general. He traveled in the highest social circles, was classically educated, and was a skilled physician, but a chill shot through him nonetheless. He was familiar with all the tales, and he knew that owls were the traditional harbingers of death.

Just before the Roman emperor Antonius died, an owl had alighted on his residence. Same thing happened to Valentinian, according to Roman histories. And before the death of the great Cesar Augustus, an owl called out.

Later in history the Italians had their revenge by consuming owls or catching them in net lures, but even in Munthe's time, and well into the twentieth century, Italian peasants traveling at night would cross themselves or touch a crucifix if they heard an owl call.

Owls fare no better in British and northern European folklore. You couldn't even mention owls in Munthe's native Sweden without putting yourself at risk of a sorcerer's charm, and killing one was sure to bring ill luck. Throughout northern Europe and even into the Near East, owls were considered the associates of witches, or dark deeds, harbingers of a death to come and were even used as ingredients in witches' brews. Shakespeare's weird sisters, for example, used an owlet's wing to strengthen their foul concoction in *Macbeth*, and later in the play, an owl—"the fatal bellman"—shrieks just before Macbeth murders Duncan. No doubt the scream of that notorious Irish herald of death, the banshee, had its origin in the wail of the little Irish screech owl.

There used to be a legend in England that the owl was in fact a pharaoh's daughter, and there was a couplet to comfort children awakened at night by the owl's scream:

I was once a king's daughter, and sat on my father's knee,
But now I'm a poor hoolet, and hide in a hollow tree.

Curiously, there are only two exceptions to this bad reputation of a perfectly innocent creature, which does inestimable good for the human community by holding down the populations of grain-eating mice.

In ancient Greece, the owl was considered a sacred bird, associated with wisdom and the goddess Athena. In fact, in some of the statues of Athena the goddess appears with an owl's head.

The other culture that appears to have a certain reverence for owls are certain tribes of American Indians. Archeologists excavating an 8,000-year-old rock shelter not far from Scratch Flat found, among the bones of more edible species, the tiny hollow bones of a screech owl. The owl could have been used for ceremonial purposes, or perhaps was even kept as a pet. In historic times there are records of pet owls kept by the Mandans in the Missouri River Valley, and the Zuni, who had a special reverence for owls, used to keep them in their houses. Small children were warned that they were all-knowing creatures. On a darker side, shamans in certain mid-Western tribes used to transform themselves into owls in order to attack their enemies, according to Ernest Ingersoll, who researched bird legends throughout the world.

None of these mystic emanations should be surprising to anyone who has ever been awakened by the shivering descent of a screech owl call at midnight just beyond the bedroom window, let alone the bizarre, strangled caterwauling of a barred owl, sounding out from a nearby wooded swamp.

• • •

By the end of March, there were a few warm days when you could smell a characteristic scent of sun-warmed grass on the stream bank. The marshes by then were alive with the calls of birds, and the nights were loud with the ringing bells of the spring peepers, calling from the nearby upland vernal pools. The woodcocks were still strutting around their little stomping grounds, peenting and flying up into the invisible sky and fluttering down again, and it was clear that finally, after the traditionally interminable New England winter, the great god sun had come around again, and spring was coming in.

CHAPTER FIVE
THE EGG MOON

By early April, toads begin to call, and the migratory birds begin returning in full force. Painted and spotted turtles emerge from hibernation and bask on exposed fallen limbs and grassy clumps, and the green shoots of the new cattails spread across the marshes. By mid-month the red maples in the surrounding swamps flare up, as do the shadbush flowers. Golden willow leaves brighten the stream-side banks, and down in the obscure, mysterious waters of the brook, stickle-backs spawn and a host of aquatic insects, snails, and worms begin to hatch out.

In the past at this season, there were vast migrations of herring, shad, and young eels in the waters of Beaver Brook. Originally this was a major event for the local native people who would gather at the falls below the brook to catch the fish and dry them for future use. But the construction of downstream dams in the eighteenth century put an end to the Beaver Brook fish migrations.

This month also marks the storied April Fool's Day, which has been a traditional day of spring pranks since the Middle Ages. The festivities actually reach farther back into prehistory as a spring celebration akin to the myths associated with the vernal equinox in ancient civilized countries. There was a Roman festival that took place around this date that was associated with the earlier Greek story of Persephone and her mother, Demeter, who was the goddess of fertility and agriculture.

According to the Greek version of the story, Persephone was out in the fields gathering flowers one day in spring when the Earth opened up and the powerful god of the Underworld, Hades, grabbed her and carried her down to his kingdom to be his bride.

Demeter heard the distant cries of Persephone as she was being abducted and later confirmed from a local peasant that she had been carried off. Demeter went out over the world carrying two torches to search for her daughter and neglected to attend to the return of spring, so the world began to dry up. Seeing this, Zeus, the brother of Hades, decreed that Persephone should be returned. But there was a tradition that if, while living in the Underworld, one were to eat a pomegranate one could never return to the upper earth. Hades fed Persephone a pomegranate to keep her, but she ate only six seeds, and so it came to pass that every spring Persephone would return to her mother and the world would flower for six months, after which she had to return to Hades. Demeter would neglect her duties in her absence and plants would wither.

By early April in North America, the rebirth of the green world is more likely to be announced by frogs. As early as February in some years the local chorus of wood frogs begins. One warm afternoon in early March I had heard the duck-like wood frog quacks calling from one of the vernal pools that dot the forest above the brook. Not long after that I heard a few plaintive calls from a single spring peeper, and then, as the nights warmed, a full-voiced chorus of peepers broke out from a shallow pool at the bottom of a field near the brook. Finally, in early April, I heard the beautiful sustained trill of toads coming from the same area, then the banjo string call of the green frog from the marshes. The bullfrogs would come later, a comfortable jug-a' rum croaking from all over the marshes.

Along with bird song, these were the veritable voices of spring; as the great almanac keeper and naturalist Donald Culross Peattie pointed out in one of his books, the call of frogs was probably the first voice of life on this planet.

Salamanders too had been showing up on wet spring roads, migrating from their underground sanctuaries to the vernal pools where they would mate and lay their eggs. Turning over logs on my way to the stream bank one day, I found a couple of the little red-backed salamanders resting comfortably in the dank cold soil, and later one afternoon in my own garden I found a rare and endangered blue-spotted salamander. It must have been on its way to a series of vernal pools northwest of the garden where spring peepers and wood frogs also occur.

This return of watery things, and especially amphibians, with their double lives, must have proved sheer miracle to the ancients and, as with so many of nature's mysteries, generated mythical explanations. One of the most bizarre is the legend that salamanders are born of fire. Except maybe for a fish or an earthworm, there is no creature on earth that is less likely to have any relationship with heat. Most salamander species spend nearly a full year underground and emerge only into the upper airs (and at that only on warm rainy nights) to migrate to vernal pools, where they mate, lay their eggs, and then return to earth. And yet the legend was described by the ancient naturalists Pliny and Aristotle and was especially popular in the Middle Ages. Among other strange legends, suits of fireproof armor reportedly were made with salamander skin in order to protect the wearer from flames.

The origin of the legend is logical. There was a detailed report of this association set down in the Renaissance by the artist Benvenuto Cellini. When he was a boy, he was watching an old burning log in the fireplace when he saw—with his own eyes—a small lizard-like creature crawl out from the burning log. He called his father to show him and to his surprise was beaten, albeit it lightly. It was a way, his father explained, to get Benvenuto to never forget this miracle of creation.

Old wet logs offer refuges for certain species of salamanders, which will crawl under the bark to hide during the day. When logs of this sort were cast into a fire, the poor salamanders would emerge—the apparent creation of the flames.

Frogs and toads are another matter. They are often associated with rain and fertility.

One cold moonlit night when he was living in Shanghai, my father was returning to school on the famous Bubble Spring Road when he heard a tumult of gongs and fireworks coming from one of the small villages. He turned off to investigate and saw a hoard of people all well bundled against the cold, setting off firecrackers, beating gongs and symbols, and chanting and kowtowing to the moon. He looked up and saw that an eclipse was in progress. He was well versed in the Shanghai dialect by that time and asked the people what was happening. They explained that a demon was devouring the moon and they were attempting to scare it away—which they did. Slowly, as he watched, the demon retreated and the full moon reappeared.

He was not sure of the use of the word "demon" and did some research when he got back to his room. The demon, he learned, was a toad. It turns out in Chinese folklore the toad is always female and, as with frogs, are associated with fertility, and often paired with the moon. But periodically the toad goddess attempts to devour the moon. She can only be chased off by ritualistic noise and chanting.

As with bears, frog and toad legends are pan-cultural. They appear in ancient Egyptian mythology as well as Meso-American and Asian lore. In Egypt, after the flooded Nile begins to recede, frogs appear in the floodplain by the thousands, along with the rebirth of vegetation, so the association with fertility seems clear. What's unclear is why in Judeo-Christian lore frogs do not fare so well.

In the Book of Revelations, unclean spirits, such as frogs, pour forth from the Apocalyptic Beast and, according to the Book of Exodus, when the pharaoh dared to question the power of the Jewish God, in order to impress the pharaoh with his power God sent ten plagues on the land, the second of which was a plague of frogs. The Nile receded, and the frogs appeared as they normally would. But they continued to increase and ultimately overran the dry land. They entered into the palace and (interestingly enough, given the traditional fertility association) even went up into the bedrooms and beds, also into the feeding troughs and ovens, and into official quarters—a veritable plague.

This story rings of legend, but as is often the case with folklore, there is historical truth behind the story. I was chatting with a friend of mine about Bear Lake in the San Bernardino Mountains in California where he and his family used to spend their summers. He told me that one year, in its wisdom, the state decided to stock the lake with trout, but there were too many blue gills and other local fish in the lake who would no doubt consume the trout fry, so the state poisoned the waters and killed all the local fish before they introduced the trout.

That summer, there was an invasion of baby toads. They hatched, hopped out of the lake, and spread across the land, thousands upon thousands of baby toads, so many that they covered the local roads, where they were run over by cars, forming a slimy mass so thick that people and cars slipped on them. Ultimately, plows had to be called in to clear the way.

It was an abject lesson in the balance of nature. The local fish would feed on the toad egg masses and thereby kept the populations at normal levels. There may have been some similar historical ecological imbalance in the Nile waters that fostered an unnatural invasion of frogs that made it into the local folklore.

The toad in Japanese folklore is also associated with water; they had the power to call down clouds and make rain. They can also exhale a sacred mist which could create beautiful illusions. In this aspect they are decidedly good spirits and the allies of holy men. But some toads are evil goblins and can create phantasms that can lure people to their destruction.

Later, in the Middle Ages in Europe, the frog's reputation is somewhat mixed. In the traditional tale of the Frog Prince, which may have Roman origins, the frog is an ugly creature, who in early versions transforms into a prince only after the spoiled princess throws him against a wall in disgust. In later versions, she is more benign and bestows a kiss upon the traditionally repulsive creature. True to the legendary fertility symbolism of the frog, he and the princess share a bed.

After the Reformation and the rise of Puritanism, frogs fall even farther from grace. The haunted grounds where the Salem witches held their Sabbaths was plagued by a host of what Cotton Mather termed "spade-foot frogs," which set up such a diabolical clamor in spring and summer only witches and demons could tolerate the place. Frogs and toads were known affiliates of the Devil.

In fact, Mather's so-called spade-foot frogs were probably spring peepers, who do create a clamor during their mating season. An amphibian called the spade-foot toad occurs farther south on Cape Cod, but it spends most of its life underground.

Negative legends of toads are various in the West, including the folkloric belief that touching a toad will give you warts—via some sort of mystical transference, one supposes—since toads do have warty skin. They are, of course, also associated with toadstools. According to folklore, they sit on top of a mushroom and in the process implant their toxins.

The mushroom they are generally depicted sitting upon (to my knowledge, no one has actually seen a toad sitting on a mushroom) is the fly agaric (*Amanita muscaria*), which is indeed toxic. It is also hallucinogenic and when consumed will induce sensations of peace and acceptance, and also flying.

According to Riggs, toads do not fare well in the Elizabethan Age, at least as far as Shakespeare is concerned: "Thou art as loathsome as a toad, ugly and venomous" appears in two of his plays.

By contrast, the Egyptian god Bes is a squat, toad-like creature who is the protector of newborn babies.

I have to wonder if there is possibly some atavistic mammalian holdover in all this that generated the poor toad's traditional reputation. When my son was about two years old, a toad leaped onto his bare foot in the garden—an act I would personally consider an honor. Even though he was innocent of any malfeasance as far as toads are concerned, and in spite of the fact that—via my encouragement—he was interested in frogs, snakes, rats, bats, and many other unpopular species, he reacted with disgust—and this from a boy who spent his first two years crawling around the garden digging up worms and whose first full sentence was "I love dirt."

Cold and clammy, covered with bumps and protuberances, and exuding strong juices when disturbed (as any country kid knows, if you pick up a toad, it will pee on you in defense), there are not many mammals that would deign to eat a toad. And only one reptile, as far as I know, eats toads—the hognosed snake, which itself is no beauty.

The warty skin of toads has given rise to a host of urban legends, some of which, as is sometimes the case, are real. One of these holds that you can get high by licking toad skin. One of the bumps on the back of some species of toads is a parotid gland that exudes a toxin when the animal is stressed. The secretions of the North American Colorado River toad, for example, contain a powerful hallucinatory substance. Legitimate researchers in the field, namely Wade Davis and Andrew Weil, in their quest for ever weirder drugs have experimented with toad licking. It was apparently a practice in ancient Mesoamerica in religious rites. Early Spanish explorers describe the practice, and evidence of the toxic substance has been found in archeological sites as far south as Chile.

Users describe fairly typical LSD-like effects, although the hallucinations are stronger and shorter-lived. Toad lickers experience a wide range of enhanced emotions, everything from abject terror to euphoria or anxiety, the loss of ego, and a sense of oneness with the universe.

Beaver Brook's native toad *Bufo americanus* offers no such qualities (as far as I know—I've never tried licking one), but they do evoke a certain other-worldly primordial visage, and even, in their own way, a certain terrible beauty, akin to the sublime "*terribilità*" evoked by high mountains and wilderness in old European cultures. Not only that, they have their own pragmatic beauty. Toads consume more than their weight every day in garden insects, most of them pests.

Ironically, in April when they start trilling, they're harder to find than they are in summer when they move upland to forests and gardens. Along Beaver Brook, they gather in shallow pools near the shore that are hard to get to and even if you could, as soon as you approach they will cease their singing. The same is true, generally, of wood frogs and spring peepers. You have to come up on them very slowly and quietly in order to see them, and even then they somehow seem to be able to sense your presence, and the pools fall silent.

Frogs are quite logically associated with rain. On rainy nights in spring and autumn in these parts you can often see them hopping across back roads in certain areas, where, unfortunately, many of them are squashed by indifferent passing vehicles. I also see them on wet mornings in the garden, and later in summer I see them arching out in acrobatic leaps from long uncut grassy areas that I purposefully don't mow.

I am actually partial to frogs. I have constructed a few frog ponds around the garden and have attempted to improve on a vernal pool on the northwest side of the property, generally to no avail. For three years I had a resident bullfrog who lived in one of the pools, and in late summer green frogs sometimes take up residence in the ponds. Generally speaking, frog populations in this immediate area are not strong. This may be symptomatic of the current worldwide decline of frogs, the cause of which researchers have been unable to definitively determine, but no doubt increased development and subsequent loss of habitat is responsible.

Tradition holds that frogs sing before rain, which seems true but not because they know it's going to rain. The moist atmosphere before rain may increase the volume of their calls, and in spring, at least, once it starts raining they are immersed in their element, so to speak, and begin to call. The French have a little children's verse that spells this out: "*Il pleut, il mouille, c'est la fête de la grenoille,*" which translates, roughly (and prosaically), to something like "it's raining, its pouring, it's the festival of frogs."

On the rare moist and sultry days in late April I can sometimes hear a beautiful, bird-like trilling coming from the forest behind my place beside the brook. It is the call of the gray-tree frog, and in my experience it does seem to predict rain.

Because of their association with water in traditions from all the world over, frogs are used as charms to bring down the rains. In a tradition vaguely similar to the processions of wren boys in the British Isles, certain tribals in central India hold frog processions to bring rain. They would tie a frog to a pole covered with green leaves and branches and proceed from house to house chanting a little ditty: "Send soon, Oh frog, the jewel of water, and ripen the wheat and the millet in the field."

Tribal people in the Americas would carve images of frogs and leave them on hilltops to bring on rain, while others would keep a frog in a pot, and punish it, if the rains failed. And African groups would caste frog fetishes into ponds and lakes and admonish them to bring on the rains.

There are also historical accounts that suggest that toads and frogs can forecast earthquakes. Somewhere I read of a Roman tradition of keeping a frog in a bowl and watching for odd behavior as a precursor to tremors, and there are stories from China and Japan of the sudden appearance of toads on the land just before an earthquake. One theory suggests that certain chemical changes take place in local waters as the eruptions begin but before they are actually experienced at the surface.

My mentor and guide in all these nature matters, Mr. Henry David Thoreau, had a genuine interest in frogs. A local farmer wrote that one morning he saw Henry standing knee-deep in a muddy pond, staring down into the water. He was still there at noon, and later in the day he was still at his post. The farmer asked him what he was looking at so intently, and Henry explained that he was studying the behavior of bullfrogs. (Little wonder that the pragmatic, hard-working local farmers thought Mr. Thoreau was a little eccentric. He graduates from Harvard, and then comes home and stands in a pond all day watching frogs.)

• • •

As the weather warmed, I began walking down to a foot bridge that crosses Beaver Brook about a half-mile downstream, probably near the spot where Tom Doublet had his fish weir. I also started taking out my small, single-person canoe and paddling through the maze of side channels and oxbows, seeking whatever adventures might present themselves. My fantasy is that someday, lost in these head-high walls of green, I will see the flashes of the surging schools of alewives of the past, or come upon a band of Indians collecting reeds and rushes for their mats. In fact, I did once come upon a group of people out in the marshes on the north side of the Great Road. They were dressed in an array of colorful shirts, and there were women in long skirts, bending over in the shallows, cutting down reeds. Of course I had to investigate. It turned out they were indeed Native Americans. They were reenactors from Plimouth Plantation, and they had—for some reason—come all the way up to Beaver Brook to get a certain species of rush that they would use for flooring in their traditional wickiups.

Mid-April is still, in our time, the season of fish migration. All along the coastal streams of the northeastern Atlantic coast that are not blocked by dams, herring, shad, and also elvers, or baby eels, start moving up to inland ponds and pools where they were originally spawned.

Had I been paddling upstream two or three hundred years ago I would have been surrounded by swimming fish. They would have come up the Merrimack River, turned up Stony Brook, and forged on against the currents, and then, having come upon the falls at the end of Beaver Brook, undeterred, they would have carried on, leaping against the silvery waterfall until they reached the quieter waters of the actual pond and stream. This spring migration must have started, eight to ten thousand years ago, and carried on until the dam for the waterworks was constructed.

I do see fish in Beaver Brook in April, but these are not the migratory fish that once crowded the stream—no shad, no herring, and no elvers. The main evidence of the local species I see are fish scales in the droppings of the local otters. And when the weather warms, I sometimes swim out to the center of the stream where there is a sandy bottom scoured out by the torrents of spring. I stand there waist-deep staring down into the murky waters à la Henry Thoreau, waiting for something to happen. Sometimes tiny dace or sunfish swim around my legs and even nip at my shins.

I have also seen what I believe are sticklebacks swimming along the banks. In April, the male sticklebacks build little nests in the shallow bankside waters to attract females, which then lay their eggs in the nest. The males then fertilize them and remain on guard until the eggs hatch.

True to poetic form, though, April can be a cruel month. There are warm days redolent of moist earth and emerging life, the air above the marshes filled with the calls of red-winged black birds, and the tree-lined shores sounding off with the whinny of flickers, the songs of purple finches, cardinals, tufted titmice, and Carolina wrens. And then the next day a high cold wind drives you to armor yourself in winter clothes and turn up your collar. Then come cold chilling rains. Then a sultry afternoon filled with frog calls, and the melodic songs of the gray tree frog. And then—snow—a dark, wind-driven burst of wet, all-encompassing snow. Until, finally, by the end of the month, a sense of true spring, with the banks of green willows flowering and an apparent assurance that winter is finally over. Osiris has risen and the world is about to burst into flower.

CHAPTER SIX

THE FLOWER MOON

It was May and all through the woods and fields the trilliums were in flower and the bloodroots were blooming. Fiddle-heads appeared in the woods above the stream bank, mayflow-ers opened, and you could see the lacework of the unfurling leaves of the maples and the hickories and the oaks.

Down on the brook, the hooded mergansers drifted past the bank, their white hoods and black bodies contrasting the pale greens and dappled button bushes of the marshes. Old Riggs Holbrook passed by one day in his battered wooden canoe, poling his way downstream like a Venetian gondolier and surrounded by flocks of red-wings and busy little marsh wrens, and the slow flight of a sole great blue heron passing

73

down the marshes. The air was filled with the scent of blooming chokeberries. Wood thrushes returned and whistled their
plaintive, fluted songs from the bank-side deciduous forest,
and tree swallows swept in above the marshes twittering and
diving and skidding across the sky.

The waters of the stream were high by mid-May and flooding the banks, and in the quiet side pools of the floodplain I
could see water-striders, swimming water-boatmen, whirligig beetles, and backswimmers. On warm evenings, mayflies
danced over the slow waters. The larvae of blackflies coated
underwater logs and rocks and the dace and the red fins began
to school. The first dragonflies hatched, and then the warblers
came back, the parulas, and ovenbirds, and the yellow-throats,
and the water thrushes, and there was birdsong—birdsong
everywhere. And it was May again and all the world around
there were celebrations.

To this day in France, people flock out to the countryside
to gather the *muguet*, the lilies of the valley, which they tie into
little bouquets and then pass through the streets, handing them
out to friends and perfect strangers, sometimes with kisses
on both cheeks. Shells and skiffs and barges pass along the
narrow rivers, some decked with flowers. And elsewhere, and
in more ancient times in Celtic lands, the traditional Beltane
fires were lit. People decorated their cattle with flowers twined
around their horns and herdsmen led them in lowing processions around one fire and then on to the next, and there were
feasts and quaffing of May wine, and dancing and torch-lit
processions. It was May and summer was a-coming in.

But not here. Not on Beaver Brook in 1700, at least. The
first English settlers who moved up from Concord and built
farms along the brook were Puritans and would have none of
it. This revelry was, after all, nothing more than a holdover
from the old heathenish, pagan traditions.

In any case, the English farmers who arrived on these banks in 1640s were hard-working pioneers with no time for frivolities. They spent May Day in the fields, plowing and planting. The first of May back then was just a spring day, like any other spring day, as if the world was nothing more than a clock, ticking the hours in one tedious procession after another. So there were no celebrations in those years. And four hundred years later, for very different reasons, the first of May also passed by without notice.

From my bankside sanctuary on the brook I could hear the hum of the tidal bore of traffic that flows eastward on the Great Road each weekday and then, eight hours later, floods back to the west, the unremarkable, boring workday over. Most of the drivers rode with their windows closed, most listening to the news or music, or chatting (illegally) on cell phones, thereby drowning out any hint of evening birdsong.

I can't say for sure, never having completed any surveys on the subject, but I would guess that the majority of the drivers do not look north or south over the great sweep of grassy marshes as they pass over Beaver Brook. They do not notice the unique shade of green that, for a matter of weeks only, appears on the young shoots of the emerging cattails at this time of year, nor the slow beat of heron wings over the open waters, nor the relaxed circle of the red-tailed hawks that often drift over the surrounding hills.

I sometimes wonder, lounging there on the dry bank, when this indifference or even actual blindness to local landscapes evolved. Henry Thoreau and his literary companions—not to mention the emerging school of American landscape painters who were active in Henry's time—were aware of the natural surroundings and knew well the names of the local streams and springs and so, it seems, were their ancestors—it was they, after all, who had given names to the various features of their new world. We are surrounded here by place names: Beaver Brook, Butter Brook, Nashobah Brook, the Nissitissit River, the Nashua River, or the old Grassy Ground River of the Massachusetts Indians, the Musketaquid. And also the Great Road, which was originally an old Indian trail named Mishimayagat in Algonquin and led from the coast to the hinterlands and intervales of the Avalon Terrane. And also Nashwatic Hill, Nashoba Hill, and, ominously, Gibbet Hill, and, farther to the southwest, the great sentinel of the region, Mount Wachusett, and farther to the northwest, Mount Watatic and Mount Monadock, which the local native people believed was underlain with silver and gold.

Somehow, over the 400-year occupation of this American land, people lost contact with their locality, and the ancient folktales and legends that tied them to the landscape and, subsequently, their place in the world. The common culprit is generally believed to be television, the internet, and social media, but the more likely culprit as far as media are concerned is the radio. Until access to radios penetrated deep into Appalachia, the local people were still singing the old English ballads, and some still spoke in the Scots/Irish phrasings and accents of their ancestors. Some of the most original forms of the early Robin Hood ballads were preserved back in the hollows of the mountains of North America and could still be recited from memory as late as the 1930s. Then came radio and slowly the old traditions were diluted.

But there were other causes: industrialization, the migration to cities from rural areas, railroads, the national highway system, and, finally, international commercial flights. The world was slowly homogenized and people lost contact with where they actually were in the world. Nature, wilderness, scenic views were somewhere else, often bounded in state and national parks, guarded by gates and entrance fees, and sometimes patrolled by armed guards, as if the natural wild forests and hills of North America were some manner of human construct like a museum or a private garden that was under the control of a government or a powerful private landholder.

The origins of place names, the stories, the legends, and the lore of the land disappeared, and the places where spirits and ghosts and goblins lurked were sanitized. There were no longer any haunted places, as epitomized in the old English verse "Over the hill and through the glen / we daresn't go ahunting, for fear of little men."

And yet, save in those places that have yet to be developed and paved over, the old stage for these little adventurous dramas still exists. There are still flowers and shrubs and trees and the shells of haunted groves. You just have go looking for them.

I know a spot near my brookside sanctuary where forget-me-nots bloom and the willow and the rowan tree—the witch's tree—grow. It is a little-known fact in our time that forget-me-nots traditionally marked the entrance to a faerie cave. One legend, probably early Christian, holds that on the third day of creation God went over the land naming all the flowers. In the process he passed over a pretty little blue flower with a yellow center. As he walked on, the little flower called out in a plaintive voice, "Forget me not," and so God so named the little forgotten flower. Another legend from Germany holds that two lovers were gathering flowers on the banks of the Danube when the man was swept away. As the current dragged him off, he called back to his lover to "forget me not."

All good stories, but the association with faeries, who traditionally could be decidedly demonic little creatures, no doubt comes out of a much earlier tradition. How the European forget-me-not came to America is not known. It's possible, but unlikely, that they were purposely transported, since they were, after all, popular in bouquets and flower lore in Europe. But it is also possible that the seeds came over mixed in ballast and then spread inland from the coasts. Either way, they are fairly common cultivated and wild flowers.

As far as the people of the brook were concerned, Tom Doublet and Mingo would not have known anything of the lore of this flower, as they are not known in West Africa and they only appeared in North American after the European invasion. But it is likely that the Caswell children did, and they may also have believed in faeries. For all their trust and devotion to the one true God, the Puritans still believed in faeries and devils and demons, witches, goblins, and howling night spirits.

Faerie lore and the belief in the presence of little people, or an alternative race of beings that shared the Earth with mortals, reaches far back into antiquity. According to Riggs, even that old dyed-in-the-wool Puritan, John Milton, wrote somewhere that millions of spiritual creatures walk the earth unseen.

Given our age of scientific advancement, it's hard to comprehend the fact that two hundred years ago, especially among rural peoples, faeries were an absolute reality. Depending on the culture and the regions where they are found, faeries go by many names and come in all varieties. The word in English derives from the French name *fee*, first Anglicized as *fay*, and then later spelled out as *faerie* and later still, and somehow sanitized and even commercialized, as *fairy*. Some, such as elves, brownies, and leprechauns were helpful—spinning and weaving and doing household chores at night, and some were evil little beings that stole children and bewitched lost travelers and seduced errant knights. As Shakespeare points out in *A Midsummer Night's Dream*, the hobgoblin Puck frights the maidens of the village and misleads night-wanderers, laughing at the harm. Seventeenth-century Puritans accepted their existence but transformed them into agents of their archenemy, the Devil. It was a known fact that they were especially common in the dark forests of New England, and they were everywhere around Salem in the late seventeenth century, the site of the famous American witch trials. Their presence, and the presence of other dark spirits in Salem, was probably enhanced by the local Barbados slave, Tituba, who shared stories and taught the two young Parrish girls, who were the primary witch accusers, all about spells and charms and her African lore of evil spirits. The girls were already primed for such beliefs, having been forced to listen to the ranting Puritan diatribes describing the terrors of Hell and the presence of the fiery worm of the Devil who flew by night, as well as other unearthly beings.

The forest was a dangerous region where, at any time of night, wildcats could mix with hyenas, goat-demons would call to one another, and witches such as the evil goddess Lilith walked the land.

Tituba may also have told them stories of the legendary Heartman, who would eat little children, or the little elf called Baccoo, who lived in a bottle and could predict the future.

The young girls knew they were in dangerous territory here—this was the sort of voodoo that was the realm of the Devil, and in time their imaginations got the better of them and they began accusing perfectly innocent, although slightly off-center, old women of being witches.

The Puritan minister Cotton Mather, who was involved in the Salem witch trials, even wrote a book about the subject of evil spirits called *Wonders of the Invisible World.* His argument was that the Devil was alive and well in Massachusetts and had sent a great company of demons do work their nefarious magic on the women—and men—of Salem.

The belief was contagious. In 1722, one of the women among the Beaver Brook community, Mary Louise Dudley, was accused of witchcraft by two young girls. She died before she was put on trial and the charges were as later refuted by one of the girls, but the acceptance of the possibility of a witch in the community was taken seriously.

This was probably made all the more real by the lore of the native people. Indians had their own devils and demons that haunted the culture. In particular there was a hideous wood monster named Hobomacho who lived in the night woods and whom the people feared. There was one report from Plymouth of a runaway slave who was forced to hide in a tree. A hunting party of Indians passed under the tree, spotted him and ran in terror. They had never seen a black man before and thought he was Hobomacho.

By the 1650s, the English and the Indians were trading goods and mixing socially and slowly beginning to understand, or at least observing, what must have seemed to both groups to be the bizarre cultural practices of their people. And one cannot blame the overly religious Puritans who settled along the banks of Beaver Brook and its environs for accepting as real the presence of evil spirits, devils, and bad faeries. The forests were indeed populated with the wild beasts and the wild men described by William Bradford. There were wolves, bears, cougars, foxes, fishers, hawks, and eagles, and larger animals such as white-tailed deer and the moose, and even woodland caribou and wood bison. And then there were the wild men, the Indians.

We have no full physical descriptions of the native people who were living around the brook in the mid-seventeenth century; the only written description we have in the local histories is of Tom Doublet, who was characterized as an old man, living out his last days on the shores of Beaver Brook. There are other accounts from the Pilgrims and a few Englishmen who had settled around Plymouth and Concord. The Native Americans were apparently well formed, and they decorated themselves with tattoos and face paint, wore elaborate hairdos, and wove feathers, skins, and furs into their clothing and headdresses. The tribe associated with Beaver Brook at this time consisted of mostly Massachusetts Indians under the direction of a powerful chief named Tahatawahn. His son, John, married a local *saunk,* or woman leader, named Wunnuweh. After John died, she married Tom Doublet and renamed herself Sara, in the Christian tradition. She lived on after the King Philip's War near the Nashobah Christian Indian village, just southeast of Beaver Brook. Artifacts, dating back 2,000 to 4,000 years belonging to her Eastern Woodland ancestors, were found just upstream from my brookside sanctuary.

If she adorned herself in the manner of her nearby tribal people, Sara probably tied her long black hair in knots and wore a decorative moose-skin skirt with a blue shawl over her shoulders and a beaded blue cloth band around her waist.

She was the last "owner" of the Nashobah Plantation land, although she may not have fully understood the legal concept of the condition. Native Americans had no real concept of property or privacy. One of the contemporary complaints about their behavior was that they would walk into any Englishman's house without announcing themselves, look around, and then openly abscond with any object that took their fancy. When it came to conversion of the natives to Christianity, one of the many laws that were promulgated in order to join the faith was that an individual would knock before entering an Englishman's house and would not "steal" (the English term, not theirs) any property.

One can imagine the shock of an innocent Puritan family sitting in front of the fire some autumn evening and having an Indian, in full regalia and with no common language, enter the doorway, range around the house, pick something up, and walk out.

One can also imagine the terror children and even adults must have felt when they heard the infernal shrieking of demons just beyond the cabin walls at two in the morning. At least they understood that these were the local demons and could be avoided or driven off by prayers and invocations to God. The Devil and his cohorts were everywhere in this terrible new world.

He was everywhere back home in East Anglia as well. Unexplained or unusual deaths or accidents among people and livestock were known to be the work of demons and human witches. All of Europe was suffering from a wave of witch hysteria at this time, and in East Anglia and Kent witch hunts were in full swing. One of the major witch hunters, a man named Matthew Hopkins, went around East Anglia uncovering witches left and right and having them tried and executed. He was responsible for the accusation of some two hundred witches in East Anglia alone. Horrid though it was, there were actually fewer witch hunts in New England in the 1640 and 1650s.

Back at home in East Anglia and Kent, where the local America colonists had come from, things were not going well. The English Civil War ended in 1651 with victory for the Parliamentarian Roundheads of Oliver Cromwell. King Charles I was executed and Oliver Cromwell took control of the Protectorate, followed by his son Richard. Theater was banished, holidays were no longer celebrated, dancing and card playing were discouraged, as was, one presumes, any vestige of excessive drinking and social gatherings. At least that was the traditional position. This was, among other sins, a period of illicit sexual license.

There is a legend that at some point in those years, the faeries left England. People actually saw them out in broad daylight. They assembled in Kent in southern England and began marching in a long migratory procession, most walking, some riding tiny faerie horses caparisoned with tinkling bells, some of them leading their little cattle, their horns twined with flowers. They wore red doublets and striped trousers, and the women wore bright striped cotton and silk gowns with slashed sleeves and caps and bonnets bedecked with feathers. They collected more and more faerie refugees as they moved northward, up around London, north through the Midlands, and into York and on to the Western Isles and the Irish Sea.

Two children, near the Burn of Eathie, in Scotland, had a good view of them and even managed to talk to them. The children spotted the cavalcade emerging from a ravine that ran through a wooded hollow near their farm. The file wound through the knolls and passed right by the young boy and girl. The boy grew curious and dared to confront them.

"What are ye, little mannie?" he asked. "And where are ye going?"

One of the faeries, who was riding a shaggy little dun-colored horse, turned in his saddle.

"Not of the race of Adam be we," he said. "We're off to Ireland. No more shall the people of peace be seen in Old England."

As is so often true with legends, there was some truth to the story. Belief in faerie lore began declining during the Age of Enlightenment and the rise of rational thought and advances in scientific exploration. But that is not to say that all the faeries left. Some stayed behind in remote parts of the British Isles, and, as the little mannie explained, many went off to Ireland, where they can still be found. Some, by whatever means, even came to North America, presumably stowed away in ballast with the forget-me-nots, the eglantine, and the Quaker ladies.

In our time, one old Irishman questioned by an ethnologist friend of mine explained that while he himself did not believe in faeries, it did not mean that they weren't there. Another older man in County Cork became upset with another American friend of mine who was visiting his distant relatives. One evening the younger group began joking about elves and the banshee, and the old grandfather grew agitated, slammed the table, and raised his voice: "You'll not be funnin' about the banshee!" he said.

It was a warning of sorts, I suppose, and he may have been worried about his own demise. The banshee only calls before a death in the family.

Another friend researching the mythic landscape of Iceland as set down in the Sagas was told more or less the same thing by a perfectly well-educated country man. He said that although he personally did not believe in trolls, that did not mean that they weren't there.

Although they were certainly living there long before they deserted England, it seems that some of the Anglo/Celtic faeries also went up to Iceland, where they were known as the *huldufulk*, the hidden people. They also were living in remote areas in Sweden, Norway, and Denmark. The *tomte* in Sweden were troublemakers. They were associated with singular places and would torment the farmers who built their houses on the faerie burial grounds. They would steal things or hide them around the house. But they could be propitiated by leaving a bowl of milk in their dooryards.

In many cultures, one of the theories of faerie origins is that they were the souls of the dead. Then, after the advent of Christianity, they were believed to be the souls of those who died before they were baptized. An alternative Christian theory on the presence of the *hulderfolk* in Scandinavia holds that God visited the home of a cottager one day, and the mother of the children was unable to wash all her children properly so she hid the dirty ones away. God was, I presume, displeased, so he said they would be forever hidden from humanity.

Here, back home on Beaver Brook, good heathen that I am, I was reluctant to dig below the stand of forget-me-nots to see if any faeries were asleep there or perhaps feasting. Had

I discovered them and joined in their festivities I may never have made it back home that day, if ever.

There are recorded interviews with country people in Cornwall that include a description of one of these feast halls.

The tale, recorded by a local folklorist, tells of a farmer who went off in search of a lost cow one evening and did not return. Searchers found his horse tied to a tree, but he was nowhere to be seen. The search crew crossed over a bog and came finally to a ruined barn. Inside they found the farmer fast asleep. He woke up, surprised to learn that three days had passed since he left. He then told them he had become lost in the bog, but at one point he heard music and followed the sound and came to a wide hall, well-lit and crowded with well-attired tiny people, quaffing and singing and dancing. The farmer drew nearer and was suddenly halted by a woman in a white dress, who warned him not to go in. He was hungry, he said, and would she get him something to eat before he left? She said she would and wandered off. There was a plum tree next to the barn, hung with ripe fruit, and he reached up to pluck it when the woman appeared again and warned him not to eat. "If you do," she said, "you'll never get back."

It was then that he recognized the woman. She had been his love five years earlier, but she had disappeared one night. People assumed she had died, but she explained to her former lover that she had followed the same general path over the bog one night and had been drawn into the faerie festival, where she ate and drank their wine. She should never have eaten. She was bound to live with them for the rest of her days.

The same thing happened, it will be recalled, to Persephone. Hades tried to get her to eat the pomegranate to keep her in the Underworld, but she ate only six seeds.

The farmer's faerie lover then led him back across a bog to the ruins of the barn and put him to sleep.

Given the fact that it was early May on Beaver Brook, there was probably no need to dig in the ground and look for faeries, at least not then, in the first week of May. They might have come up to the normal plain of reality quite on their own. Proof of their presence can be found in the circles of faerie ring mushrooms. The circular growth of these tiny mushrooms offers the evidence of the nighttime ring dances of the local faeries.

There are certain seasonal changes that allow the unseen, supernatural world to break through the normal barriers of time and space and cavort over the land. One year on Walpurgis Night, which is the night before the May Day celebrations begin, I took my life in my hands and after dark went down to the bank. On this night, it is—or was—a known fact that witches would take flight, searching for victims. This was an uncommonly warm and close evening, more like summer than spring, and the air was still and close, and you could smell frogs and fecundity. I sat for a while, watching the fog-shrouded moon for crossing witches (and also migrated birds) but although I kept an open mind, nothing appeared, neither bird nor witch.

I may not have encountered any supernatural beings that night, but I could understand why the eerie sounds coming off the night marsh could be unsettling for one not accustomed to the noises of the night world. The presence of supernatural wild things would indeed seem possible. I've heard the bark of foxes on the far shore, the booming hoot of barred owls and great horned owls, and the howling of coyote packs at these times. The plashes, grunts, croaks, whistles, bird chirps, growls, and quocks echoing over the marshes offered a potentially unsettled landscape of ghostly presences. But I recognized most of sounds—the green frog banjo-like "ploncks," the rumbled jug-a-rum croaks of bullfrogs, the comb-like scrapings of the leopard frogs, and a periodic "quock" from night herons. Also, in warmer months, insect calls: meadow crickets, field crickets, katydids, and snowy tree crickets. Father downstream one night, in a cove surrounded by dead trees I heard an odd-grated scratching sound that took me a long time to identify. It was coming from inside the trunk of one of the dead trees. I was a little worried, thinking it might be the ticking of the death watch beetle, which, like the banshee, foretells a death, but I finally determined that it was a long-horned wood boring beetle.

Once there by the stream bank at night I did hear an unearthly yowling scream sounding out from the dark forest behind me. This was no bird call, or fox yowl, nor a coyote or a rabbit's screaming death throes, nor was it the strange whinny bark of a deer. It was unlike any sound I had ever heard in the local woods and swamps. It was the cry of a maniacal spirit woman, a doomed soul calling from the Underworld. The ghostly descent of the call of the screech owl is one thing, but this call was truly demonic, beginning as a low clarinet-like sound and then rising and falling and wailing and chattering. I never did figure out what it was, but Riggs Holbrook told me that long-eared owls can sometimes create bizarre screaming tirades.

I think I might have preferred having a demon here amidst the encroaching sprawl of suburbia. He or she might have been able to hold at bay the course of development that is overwhelming my streamside fantasy land.

To new-coming settlers from England, all these barks and screeches and yowlings would make a good argument for the presence of the Devil and his company of demons in New England—proof that the Puritans had settled in the country of the Devil himself and had to practice their faith all the more assiduously to hold him at bay.

However, not all who settled in this region in the mid-seventeenth century were of the Puritan faith. There were a few Englishmen around who were comfortable with the land and the forest, and also with the local wild men. William

Blackstone, the first resident of the Shawmut Peninsula, where Boston is now located, was a somewhat eccentric Anglican minister who lived alone on what is now Boston Common with a herd of pigs and a library of books. He got along with the local Indians and presumably did not fear the calls of the local wild birds and mammals, let alone the demonic spring peepers that still call on spring evenings around Boston Common. Not long after the Puritans arrived, in the 1640s, he decamped for Rhode Island, saying that he was more oppressed by his Puritan neighbors than by the king.

There was another contemporary apostate who had set up a trading post just south of Boston and close to the Plymouth Plantation, which was established in 1620. Thomas Morton was also an Anglican who, unlike his Plymouth neighbors, was living the high life among the Indians, eating well, drinking and carousing with the Indian maidens, and trading. One year in May, he and his merry company of equally irreligious men set up a maypole and invited the native people to join the festivities. Down in Plymouth, William Bradford paid a visit and recorded the events that followed. He was suitably horrified by the "beastly practices of ye mad Bacchanalians" he encountered there. He found Morton and company in a state of inebriation, frolicking around the maypole with the scantily clad Indian maidens "like so many faeries, or rather Furies," dancing, and drinking, and singing "——and worse practices," he wrote.

This was too much for the sanctimonious Pilgrims. Bradford sent Myles Standish up to the post with a company of soldiers to imprison Morton and send him back to England. There was a little confrontation, but Morton's followers were all drunk and he was easily apprehended

The Pilgrims were right, of course, to be disturbed by the proximity of a maypole. The rituals associated with maypole dancing and the whole body of the early May festivals, from Walpurgis night in Germany to May 1, international socialist celebrations were decidedly pagan in origin.

The Romans may have started the tradition with the celebration of the goddess Floralia, a six-day festival of dancing and drinking, a celebration of spring flowering. The festival had lewd elements, prostitutes were invited, and there was naked dancing around the monument to the goddess Flora. The eighteenth-century artist Giovanni Tiepolo has a huge colorful painting of the arrival of Flora, now in the Fine Arts Museum of San Francisco. Bare-breasted Flora lounges with a haughty though seductive look in a flower-bedecked chariot drawn by winged putti and accompanied by half-clothed maidens.

Later in history, or perhaps contemporaneously, Germanic tribes held ritual dances around sacred trees in this season, later transformed into the traditional maypole. Scholars have read all sorts of symbolism into the maypole celebrations, among them Freud who saw, perhaps needless to say, the pole as a phallic symbol. And in fact there are other interpretations that the celebration was a spring fertility ritual designed to encourage crops and livestock. The tradition goes back even further into rituals of tree worship among the people of the great Ciminian Forest that once covered Europe. The sacred trees and poles were decorated with flowers, and the celebrants would dance and parade around them singing and chanting ritual phrases.

As the twentieth century passed and the electronic age of the anthropogenic twenty-first century advanced, faeries were driven off into even more remote spots on earth. The old country people of Ireland have also left us, carrying with them the living evidence of the little people.

And yet, who knows? Just because we don't believe in them anymore doesn't mean they're not there.

CHAPTER SEVEN

THE STRAWBERRY MOON

On a fine afternoon in June in the year 1859, Mr. Henry David Thoreau went out for a walk. He was living at home at his parents' house in Concord that year, and he set out walking southeast through the woods around Walden Pond where, in the late eighteenth century, a family of free Blacks headed by Sippio and Fenda Freeman maintained a small holding and a farm shortly after the Revolution. Thoreau had visited the site of their holding in the past and reported in *Walden* that he enjoyed fruits planted by Sippio and had also visited his grave in Lincoln. His epitaph stated that he was a man of color, "as if," Henry wrote, "he had been discolored."

There was another African American living near Walden named Zilpah White who maintained a small plot of land with a cabin and kept a flock of chickens and wove baskets, brooms, and chair seats for the Concord people. She was a former slave, an independent woman who had been freed, and she could have found work as a maid for any of the Concord households but decided instead to make it on her own. Her diet was decidedly limited; she lived on eggs and the produce of her garden, and she kept a cauldron of beans and peas boiling over a fire in front of her shanty. Thoreau imagined her there stirring her pot, "witch-like."

He may have picked up this identification from the older locals of Concord. Single women of color were often suspected of witchcraft. A free African American, Brister Freeman's wife, was identified as a "fortune teller"—in other words, an exotic outcast with magical but suspect abilities. She may have been considered what was called a "cunning woman" in early eighteenth-century England. These were often single women who could perform magic, cast spells, and predict the future. They were akin to witches but were not necessarily evil, and even though they were outcasts, they were not persecuted.

Vandals burned Zilpah White's house down when she was in her mid-seventies, killing her chickens and her dog and cat. Zilpah's suspected witchery may have been the motive. But she was nothing if not independent minded, and she rebuilt the cabin and lived there until she died, at age 82.

Henry Thoreau's life in the woods was hardly the wilderness sojourn that some readers consider it to have been. The whole wooded area around Walden Pond was a shanty town. The tract had a small settlement of huts where former slaves, Irish, and outcast local drunks maintained their shelters. Henry used to poke around the woods of Walden looking for cellar holes and the ruins of past lives such as that of Zilpah White and Brister Freeman. Unlike the good citizens of Concord, Henry had developed an appreciation of the outcast lives of these people, and during his time at Walden he befriended a few of his fellow woods dwellers, in particular, the French Canadian woodcutter Alex Theirin.

As he walked southeastward toward Flint's Pond on that June day, Henry noted that the red maple seeds were partly blown away by summer breezes. On the way he came upon a small sixteen-inch striped snake that he examined closely and identified as what we would call in our time a garter snake. He wandered on and found that the strawberries were beginning to ripen on some of the cleared hills. And then, approaching the pond, he found the nest of a rose-breasted grosbeak. This, too, he examined carefully, noting the height and the position of the nest in a thicket of catbrier and high-bush blueberry. Soon enough he heard the parents squeaking an alarm call, and then as he approached the nest he saw the male and noticed that there was only one egg. Thoreau, as well as other naturalists of the time and also boys, would commonly collect the eggs of birds, but Henry left this one where it was and carried on.

Later in that first week he walked out again, ranging through the Walden woods and beyond. All along the way he recorded in scientific detail the structure and status of the local wildflowers.

Naturalists in those days were interested in phenomenology, the study of the dates of bloom times and other seasonal events. One hundred and fifty years later, Henry's records became an accurate baseline. Recent studies of the flowering times of the wildflowers have provided useful records of the advance of global climate change. Flowers around Walden and what is now known as Thoreau Country are now blooming two weeks earlier.

There is something about the region west of Boston that seems to have attracted naturalists. Even before Thoreau's time, scientists were scouring the area, recording the bird life and the rivers and the plants and animals. The tradition continues to this day, thanks in part to the presence of a tract of land known as Estabrook Woods, part of which is owned by Harvard University as a research field station.

In late spring, in a throwback to the ancient festivals of spring, a local Concord art center holds a celebration of the coming of the growing season and the great mother Earth and the life of the river the Indians called Musketaquid. Children and adults spend the winter constructing huge papier-maché floats of local animals and plants and then hold a procession through the streets, accompanied by drummers and local bands. The event is preceded by a river ceremony in which a green and blue clad water nymph, draped in water weeds, a veritable spirit of the river, rises from the depths and delivers an oration on the beauty of nature and the running waters of the Earth. This is followed by more drumming and ritual choruses, and then the parade marches off. The procession ends in front of the local art center, where there is more music, as well as tables with displays and environmental literature and food.

Celebrations of a related sort also take place on the marshes of Beaver Brook at this time of year. Dawn comes early in June, and over the course of a few years I got in the habit of canoeing the course of Beaver Brook to review the festivities. I often miss the sunrise, which takes place around 4:30 in the morning during the month, but I manage to get out on the marshes early enough to hear the full dawn chorus spilling over the marshes from the surrounding woods. In contrast to the Musketaquid Festival, with its marching bands and troupes of drummers and dancers, this is a seemingly more poetic celebration, featuring birdsong as a backdrop, but it has more fertile, even erotic overtones. All the birds and the frogs and the insects are singing for sex at this time of year.

Earliest to sound off are the robins. I can hear them before I even wake up, singing in the garden. But down on the marshes the loudest sounds are the near constant harangues of the male red-winged blackbirds, of which there are many along in this section of the brook. I can also hear the distant whinny of flickers, the whispers, cheeps, squeaks, whines, and buzzing of the warblers, the songs of the wood thrushes and veeries, wood peewees, and all the other passerine birds who move through the region between April and June in wave after wave of migrants.

As if to complement the chorus, the very air at this time of year is redolent of fertility—the fresh, dank odors of the woods, the rank, marshy smell of the banks, and the watery smell of the brook is almost intoxicating. I sometimes simply lie in the bottom of the canoe and let the boat drift until it fetches up on a bend. I'm in no hurry on these trips, and I sometimes allow the canoe to run aground and just sit there, doing nothing.

I remember in particular one hot June morning trip. I had put the boat in by the Great Road bridge and paddled slowly downstream, the warm sun on my right cheek. About a half-mile downstream I came to the bank where I normally spend my mornings and rested there for a while, taking in the familiar scene of the freshly leaved-out trees on the eastern bank before pushing on. The waters were still high, and in some sections, in particular at the narrow site where Tom Doublet's father maintained his weir, they run faster. Another half-mile or so to the north I came to the local beaver dam, which causes the waters to slow and rise up to the dry banks, making it easier to portage around the dam. Downstream the brook narrows again and follows its normal twisting, snake-like route.

June is the season of reptiles and amphibians, and all along the shores that day green frogs and bull frogs showed them-selves, some sitting on the banks and flats, some poking their noses up between the water lilies, eyeing me as I approached, only to duck under at the last minute when the boat drew too close. Turtles, too, lined the banks, mostly painted turtles, but also the occasional spotted turtle, and at one point I saw what appeared to be a huge flat rock drifting across the stream just under the bows. It was one of the local snapping turtles, some of whom in this marsh grow into Pleistocene-sized monsters.

One year in June, I took my children out on this route specifically to count turtles. In the three-mile stretch of the brook between the Great Road and the lower bridge, we counted over 350, mostly painted turtles. We also saw a lot of snakes, generally black water snakes, but at one point we saw, swimming downstream just ahead of the boat, a narrow ribbon snake.

Another half-mile below the beaver dam, the stream narrows again and runs under a footbridge, where I also spend a lot of time in warmer seasons. There is a trail from the bridge that winds though open land and old fields, and from spring through to late autumn there is a host of wildflowers, each blooming in its season. I often walk down to the bridge from my house, and whenever I get there by boat (usually by paddling upstream) I get out and take a short walk.

Most of the plants in these meadows are common open-ground flowers such as Quaker ladies, daisies, hawkweeds, dandelions, and asters. Nowadays most people who know flowers would be able to name any of these—they are not by any means rare plants, but they do make up the colorful palette of the landscape there.

I have a running argument with an amateur botanist friend of mine who insists, whenever we are discussing a certain plant, that I use the scientific name in the Linnaean classification system. I agree with her when it comes to technical descriptions in scientific journals, but in everyday English conversations my argument is that these common names speak of the erstwhile human connections with nature. Each flower carries a deep folkloric history that reaches far back before the Linnaean system, which was developed by the Swedish botanist Carl Linnaeus in the mid-1700s. Not only that, I love the sound and the connotation and the historical resonance of country life implied by the names—they speak of sunny days and summer meadows: Queen Anne's lace, black-eyed Susan, sweet Annie, ragged Robin, daisy, Quaker ladies, sweet clover, mayflower, and all the rest.

Some of these common names have ominous overtones, as with, for example, aconite, or monkshood. This handsome blue flower, which has been cultivated as an ornamental garden flower, is highly toxic and was used by retreating armies to poison wells, as well as people, and also wolves—which is the root of another popular name for this plant, wolf's bane.

The opening line of John Keats's poem "Ode to Melancholy" refers to the plant as a convenient means of forgetfulness—or, as some critics, such as Riggs, suggest—suicide: "No, no, go not to Lethe, neither twist / Wolf's bane, tight-rooted, for its poisonous wine."

In connection with this dark aspect of popular flowers, there is a group of plants, many of which grow along the banks of Beaver Brook and its surrounds, whose popular names relate to witches, such as witch hazel or witch's broom. And there are many more plants, most in the nightshade family, that were purportedly used by witches either to poison people or cast spells. One of the more interesting of these is Jimson weed, which grows upland from the marshes near one of the Scratch Flat farms. It puts out a beautiful large white flower, the subject, incidentally, of one of Georgia O'Keeffe's large flower paintings. But it is highly toxic, and ingestion, or even apparently repeated contact if untreated, can kill you. One theory holds that Elizabeth Parris and Abigail Williams, the girls of Salem who first began exhibiting signs of possession by the Devil, had somehow come into contact with Jimson weed. Symptoms of poisoning include hallucinations, tingling feelings, extreme agitation, and seizures, which were all the symptoms of possession exhibited by the girls.

One of the other symptoms is a sensation of flying, and there is a related theory that the witches of Europe would purposely intoxicate themselves in order to fly off to the supposed witch's Sabbaths.

But that's only the dark side of wildflowers: most of the other lore involves an association with joy and beauty, love, and also—of course—faeries. Flowers figure heavily in faerie lore. Shakespeare's *A Midsummer Night's Dream*, obviously, takes place in June when in faerie England all the world was abloom with bluebells and cowslips, eglantine, and rose. The queen of the faeries, Titania, sleeps in a bower overgrown with wild thyme and woodbine; Puck frequently refers to flowers in his travels, and Oberon, the king of the faeries, also mentions flowers. All these plants and their associations would have been known to Shakespeare's audiences, as would the realistic portrayal of the hidden world of faeries. Faeries and elves were very much a part of the lore of the Warwickshire countryside.

Flowers are also very much a part of Greek and Roman mythology. Half the wildflowers that grow in Europe and, after colonization, in North America seem to bear Greek scientific names and are associated with incidents in mythologies. The red anemone sprouted from the blood of Adonis, for example, and a crocus grew from the drops of blood of Prometheus, who was chained to a rock while an eagle tore out his liver as punishment by Zeus for giving fire to mortals.

In order to check out what was in bloom in the meadow the morning of my trip, I landed the canoe on the eastern bank and took a short walk through the meadows. This was a clear day after a night rain, and all the grasses and flowers were sparkling with dew. I am not sure what time of day it was, but as I turned a bend in the path I came upon a handsome young couple, for some unfathomable reason, formally dressed. He was wearing an off-white summer suit, with a white shirt and a red tie, and she was dressed in formal evening clothes, a subdued yellow gown with a low-cut décolletage and a ringlet of flowers in her light-colored hair and a black ribbon around her neck.

I spread my arms in faux surprise when I saw them, and the young man, somewhat apologetically, showed me his camera.

"We're just out taking prom pictures," he explained.

"Too bad," I said. "I thought you were just walking home from a grand ball, or perhaps your wedding."

They laughed politely and glanced at each other and then explained that they liked the light at that time of day and so they were out taking pictures for their upcoming prom that night.

"He's a photographer," his lady friend said.

"Sorry," the boy said, as if they were out on some illegal venture.

"Why be sorry?" I asked. "Everyone should be out on such a morning as this."

They laughed again politely and glanced at each other, which made me wonder if perchance they had other plans once the photo shoot was completed.

No surprise, June is the traditional season of marriages, proms, and grand balls. The frogs and toads, birds, and bees know that. So do the turtles and the snakes.

I wished the couple luck and wandered on.

The waters were so high that day I wasn't sure I could make it under the bridge, so I hauled the boat over to the downstream side and relaunched. The marshes widen at that point, and there are many more bends; paddling on, in time I could see the houses set on a rise on the western banks. You would think the main Beaver Brook bridge was just ahead, but in fact to reach the takeout spot there are so many bends it takes another twenty minutes or so of paddling to reach the end.

After many a turn, I finally landed the boat, hauled it into the brush, hoisted the paddles on my shoulder, and walked home.

• • •

Along with flowering plants and bird song, June is also—at least in these parts—a season of rains. On June 17, for example, Thoreau records that there was a steady heavy downpour. On Beaver Brook, one hundred and fifty years later, there was also a heavy rain. In fact, there was a week of heavy rain and everywhere the waters were rising. The three rivers of Concord overflowed their banks and ran out over the streamside fields and lawns and the floodplain forests. Cellars were flooded, certain roads were closed—and still fell the rains.

The waters of Beaver Brook rose as well, but the wide cat-tail marshes that surround this section of the brook absorbed most of the excess, although the waters did run over the nearby Great Road, slowing traffic.

Finally, after nearly eight days of steady rain, the skies slowly cleared. But of course, as is the case in floods, that did not stop the rising rivers; it took a few more days for the waters to begin to drop. I took another, very easy run down the length of the brook that year, skimming over sections of the brook I would never have been able to get to in a normal June day.

Floods are a natural phenomenon; they are only a problem for real estate and people, and damaging floods have been made worse by destruction of forests in upstream watersheds. In Henry Thoreau's time, the causes of flooding were not clearly understood. It only became clear after the Civil War, when the American ambassador to Rome, George Perkins

Marsh, who had an interest in forestry, began studying the history of the Italian landscape. He determined that the cause of the devastating floods of the Arno, in Florence, was the fact that the trees of the upland hills and mountains had been stripped of trees, which had formerly absorbed the waters of heavy rains. Marsh expanded the concept in a book called *Man and Nature,* which laid out his theory that human intervention with natural systems was the cause of natural disasters such as soil erosion, flooding, and mudslides. The book took decades to gain some recognition but nowadays is considered a seminal work in the science of ecology. The protection of trees and shrubs in the watersheds of rivers and streams is now an established method of flood control—although not always practiced, obviously.

For different natural reasons, there were historic and prehistoric floods long before the newly evolved Neolithic cultures began cutting down trees for buildings and agriculture. One of these causes was the period of Pleistocene global warming and the subsequent melting of the glaciers, which generated worldwide sea level rise.

We are now living at the end of a postglacial warming period, a time when, theoretically, the climate should be cooling. But as we know (or at least as most of us know), the world climate is changing and the ice caps at the uttermost ends of the earth are melting. Whole islands are sinking beneath the sea and coastlines are receding. And it is all happening at an unprecedented speed.

The record of the last great sea rise is fixed in the geological strata and can be read by geologists and botanists. But it is also true that the last great flooding occurred within the era of human consciousness. And although the events associated with the flood took place in prehistoric times, before the advent of agriculture and the invention of the written word, that does not mean that the record of its coming and goings was lost to history. The great floods of the pre-literate era were recounted in the folklore of the world.

The story of a massive flood that covered all the earth and drowned most of the world's living things appears in the folk-loric creation histories of a wide variety of cultures.

One of the oldest extant cultures on earth, the Aboriginal Australians, who were settled in Australia some forty or fifty thousand years ago, have any number of flood tales, ranging from the story of a primordial snake who called for rain and caused the waters of the world to rise, to folktales in which, for various reasons, rains fell for a long time, until there was no dry land, and all the people drowned.

The Tinguian people from Luzon, Philippines, say that when the god Kaboniyan was unhappy with his people, he sent a flood to cover the earth. But Fire hid itself deep inside a bamboo tree, along with stone and iron. Men later learned how to retrieve it from these hiding places and rekindled the earth. Another ancient preliterate culture, the Bahnar, who lived near Cochin, China, have a story in which a vengeful crab caused the sea and rivers to swell until the waters reached the sky. The only survivors were a brother and sister who took a pair of all kinds of animals with them in a huge chest. They floated for seven days and nights. Then the brother heard a cock crowing outside, sent by the spirits to signal that the flood had abated. All disembarked, birds first, then the mammals, and then the two people.

The Hindu have a story in which Manu, the first human, rescued a small fish. Later the fish warned Manu of a coming great deluge and told him to build a ship. When the flood waters rose, the fish came back and led Manu to a northern mountain and told him to tie the ship's line to a tree to prevent it from drifting. Manu, alone of all creatures, survived. He made offerings of clarified butter, sour milk, whey, and curds.

From these, a woman arose calling herself Manu's daughter and the earth was saved.

There is another famous flood that was actually recorded in writing: the Sumerian tale of *Gilgamesh*, set down some 9,000 years ago. And then 4,000 years ago, the Hebrews told the same classic flood rescue story in the tale of Noah and the Ark.

And so it goes, down through the ages, rising waters, a few ethical, wise human beings or gods who save the world for the future.

The point is, there is nothing unnatural about floods. The problem seems to be that for all our histories and record keeping, human beings seem to have short memories and as a result tend to settle in floodplains and on coastal shores prone to flooding. Then they return as soon as the flood waters recede, trusting in nothing more than suspect technologies to save them from the coming deluge, which according to the most recent models will come into full force by 2030, inundating most of the coastal cities around the globe as well as all the low-lying islands.

• • •

I walked down to the footbridge on one of those rainy June mornings and went out over the brook and leaned on the rail. A lone goose was calling unseen from somewhere overhead—a bell ringing in an empty sky—and I lingered on the bridge, staring down into the dark swirling waters of the brook. The stream was full and brimming into the marsh grasses, and it funneled itself down in wandering, dark, wavering lines and then channeled under the bridge and bubbled over some moss-covered rocks on the north side of the bridge in chopped silvery waves.

Staring down into the smooth waters in the steady rain, I imagined groups of people pouring over London Bridge, making their way with black umbrellas, or more colorful images of Japanese people with spiky bamboo umbrellas that I had seen in my father's collections of Ukiyo'e woodblock prints. A line from some obscure Chinese poem my father had translated came to mind: "Six bridges cross the West Lake scene, with willow and peach trees in between."

Other watery images and lines of verse and visions of waters came to mind, the clashing rock of Charybdis, Jason and his Argonauts rowing up the clear river Askanios, the waters of Babylon, and the still waters of the Psalms, and on the heels of that, Henry Thoreau's journal entry that "time is but the stream I go afishing in."

A river, no matter how large or small, is really not any one thing. It is a compilation of waters, and the waters are a compilation of lands, of hills, dells, swamps, upland marshes, forests, bogs, and those mossy little sinks you come across on mountaintops where wood frogs and toads seem to congregate. The essence of a river, or a wide stream such as Beaver Brook, is not what you see; it lies somewhere in the surrounding hills, between water and sky, and the narrow summer banks and the wide flooded shores of spring. And the meaning of the river, in the larger sense, is obscure at best. You have to have lived through a series of years in one place to know that. Henry Thoreau, who some would say assumed the wisdom of age before he died at age 44, said that if you can know the local waters you can know the universe. He ranked "our muddy and abused Concord River" with the great rivers of the world— the Mississippi, the Ganges, the Nile. He saw the river as a constant lure to distant enterprise and adventure, an invitation to explore the interior of continents. "Dwellers at headwaters would naturally be inclined to follow in the trail of their waters to see the end of the matter," Henry wrote.

He was thinking of earthly territory, of course, and the sea, but, as always with Thoreau, he was also thinking of the great transcendental metaphors that are embodied in the natural world. "What a piece of wonder a river is," he wrote. It is the natural conclusion for anyone who takes the idea of river to the uttermost ends of the earth.

But in the end, it may not necessarily be age that allows insight. I once knew a little boy who from an early age had a natural fascination with running water. One day, standing on a bridge above the roaring waters of a brook, he turned, spread his arms wide, and slowly drew them together and announced to no one in particular, "All the waters of the world come together."

I remembered that line for years.

. . .

Walking home that day from the bank, "profitably soaked" as Henry said of himself after one of his rainy sojourns, it occurred to me that this was June 21, the day of the summer solstice, which along with December 21 is one of the most import festival days in the common stream of myth.

Normally on this day Kata and a group of friends from the art center would light a bonfire and have a big fire-jumping party, fueled with traditional mead and wine. I called her when I got home but she said the party had been postponed. The rain that was currently streaming down was supposed to continue into the night and only end later the following day.

No fires planned, she said.

The custom of lighting midsummer fires and fire leaping has deep roots in agricultural societies and is often associated with a means of increasing the fertility of the fields. In some regions the ashes of the midsummer bonfires are spread on the fields to help in the next harvest. But fertility was not the only benefit. In some festivals, the fires were fed with pungent herbs, often mugwort, and the thick smoke would repel dragons. The fire rituals also kept demons and witches at bay and also any of the evil spirits that might pollute or cause crop failures. The fires would also protect cattle; herds were driven between two separated bonfires or through the ashes once the fires died down. Unlike the winter solstice, in which the rituals were carried out to assure the return of the sun, the midsummer fires were celebratory festivals with dancing and music, and couples sometimes leaping through the fire hand in hand—a practice that was encouraged by Kata and her crowd.

Not that there were, at least not in primordial eras, some obvious astronomical dangers at the time of the summer solstice. The sun, after all, was beginning its long decline to winter, and some of the traditional fire ceremonies reflected this. In certain districts, a huge straw wheel was constructed and set on a hill above a body of water. At a signal, the fiery wheel was lit and encouraged, sometimes with the use of an axle guided by two strong boys, to roll down the hill and extinguish itself in the waters.

The Lithuanians, whose pagan culture, by the way, was the only one in which the sun god was female and who were the last European group to be Christianized, had a solstice ritual that endured even into the twentieth century. On the day of the solstice, adults and children would create large wreaths that they decorated with colorful ribbons and fixed with a candle in the center. These they would launch into lakes or ponds after sundown. Bonfires were set ablaze and the people stayed up all night leaping through the fire, singing and dancing. At the stroke of midnight, they would break up into small groups and go off into the forests to search for the *Paparatu Ziedai,* which is the "flower" of the fern and is believed to bloom briefly at midnight, which explains why they were rarely found. (Ferns, of course, are non-flowering plants.) Just before dawn, people would go swimming, since it was believed that on the night of the summer solstice, water has healing powers.

In the Christian era, the Church associated the holiday with St. John, and priests and even bishops would participate in celebrations, but the pagan rites were unchanged. One of the more interesting and elaborate of the summer solstice rituals also lasted into the twentieth century in Normandy. A group known as the brotherhood of the Green Wolf (an obvious reference to ancient vegetative spirits) would choose a new chief for the year. The wolf would dress himself in a green gown with a conical brimless green hat and lead a procession with a crucifix and choir to the parish church, where a mass was said and then at sundown the fires were lit. People would collect by the fires while ringing hand bells, and a young man and a maid bedecked in flowers would begin to dance around the fire. Soon they would be joined by members of the brotherhood, all clad in green robes, and the assembled group would circle the fire hand in hand. In time, the brothers would select a certain participant for the next Green Wolf, who would then flee, and there followed a wild chase as the brothers tried to catch him. He would defend himself with a flowered staff, and once he was caught the group would pretend to throw him into the fire. Following this the celebrants would retire for a ritual meal.

This was—more or less—a Christian religious festival. But at midnight the festival would change course and become decidedly pagan and licentious, a veritable Bacchanalian celebration. Christian hymns would be replaced with salacious, bawdy songs, the fiddlers would run wild, and there was dancing and drinking and (as that old pilgrim William Bradford phrased it) "worse practices." The next day the party would continue with parades, and bakers would create a huge bread loaf pyramid, which was decorated with ribbons. Finally, the hand bells were returned to the church and the new Green Wolf would assume his position.

• • •

The rainy month of June with all its traditional flower and solar festivals finally ended in sun. The succession of cloudy days faded, the god of the sun showed his face, the flowers of the fields bloomed, nestling robins and other fledglings began to take wing, and all things that love the sun were out of doors. Summer, finally, had arrived.

CHAPTER EIGHT

THE BUCK MOON

O n a hot morning in early July, I went to down to the brook only to find that my place on the stream bank was occupied. A large northern water snake lay coiled on the shore. I was loathe to bother him, but on the other hand I was looking forward to an hour or so alone by the brook, so I decided to sit nearby in the shade a little higher up on the bank and watch the snake, hoping it would eventually uncoil itself and slip into the waters.

Water snakes are one of the few aggressive snakes in this region. If disturbed they will attack rather than slither away as do garter snakes and milk snakes. In fact, I was bitten by a water snake years ago at a nature center where I used to work, a place where, incidentally, I had been bitten by any number of animals—squirrels, meadow mice, many harmless snakes, and even a possum, whose jaws are so weak its bite did not even break the skin.

This was, withal, a very large and beautiful snake. It was coiled with its head resting above its tail and was patterned with reddish cross bands and splotchy markings on its side. The coloring was faint, though, and I think this must have been an older snake since the colors of this species tend to fade to black with age. Also, this was a decidedly fat snake; it looked a lot like a venomous water moccasin save that it had a narrow head and not the wide triangular head of its southern cousin. Water moccasins do not occur north of the Mason-Dixon Line, in any case.

That particular summer seemed to me to be unusually filled with snake adventures. Earlier in the season, with a resident herpetologist, I had made an expedition to Mount Tom on the Connecticut River to see a den or hibernacula of the rare eastern timber rattlesnakes. It was a good day for snakes—no wind and the spring sun was warm, perfect conditions for early emerging snakes that spend the winter denned together in rocks, on south- or southeast-facing slopes, and there was a group of four or five handsome rattlers basking beyond their den.

Then late in June at dusk, driving north on a back road close to Split Rock Point on the western banks of Lake Champlain, I had seen a huge dead snake stretched out beside the road. My wife and I were headed for a dinner in Essex, and it took me a minute to realize that what I had seen was no ordinary snake—it was much larger than any of the other snakes that live in the area. But we were late and I thought I would check out the species when we returned later that night to Westport. But only a few hours later, when we drove back, the body was no longer there. I thought maybe a raccoon or a perhaps a bobcat had dragged it off the road and eaten it, and we went back the next morning to see if there were any remains in the nearby woods. As we were wandering around searching the brush, a truck pulled up and a young man joined us. "You looking for the snake?" he asked.

It turned out the night before he too had seen the snake and had picked it up and taken it home. He said it was indeed a timber rattlesnake, as I had thought, and was rare in that part of the Adirondacks, but that there was a colony near Split Rock. He explained that he was an amateur herpetologist and intended to report his find to the state authorities since this was listed as a rare and endangered species in this region.

My third snake encounter that year was the most significant as far as myth and legend are concerned. I actually saw a contemporary version of an ancient Minoan snake goddess. I was walking back from the bridge one afternoon when I saw a young woman coming down the trail holding something close to her chest. As I drew near, I saw that she was carrying a very large garter snake that she had recently caught. The snake seemed relatively calm, and she had draped it around her neck and was walking along indifferently as if it were perfectly natural, even here in the outer suburbs of Boston, to take a walk through the meadows with a snake wrapped around one's neck.

We might have greeted one another and passed on. But of course I wanted to talk about snakes. Who wouldn't? This young snake goddess was uniquely beautiful, green-eyed with black hair that she had trussed up behind her head with two coiled braids. She reminded me of a performance-art event I had seen at a gallery opening a few years earlier, involving a performance in which another snake goddess danced around a fire with a medium-sized boa constrictor twirled around her writhing arms, thereby associating herself, I suppose, with the deep historical elements of fire and the primordial Earth Mother snake.

I asked this particular snake goddess about her snake, and she explained that she had caught it earlier and was taking it down to the brook to free it. "I believe it's a ribbon snake," she said, "and they prefer water."

We chatted for a while about local snake populations, and she explained that she was an amateur naturalist and was always finding snakes. "I love snakes," she said. "Also turtles."

I picked up on this and asked her if she knew the old Native American legend of Turtle Island. She did, and she also knew about other local Native American animal tales and creation myths.

I then asked if she knew about the ancient associations of snakes with female goddesses. Archeological digs in Crete in the nineteenth century had uncovered figurines of women holding snakes in each hand. The artifacts were found in household sites, and it is believed that the figurines were associated with snake cults and represented female household deities.

She said she didn't know anything about Crete and wasn't aware of that connection, so at the risk of seeming decidedly weird, I explained as succinctly as I could the deep association with chthonic Magna Mater, the pre-Hellenic Earth Mother, who was deeply connected to snakes. She was the chief deity of southern Europe and Asia Minor until she was replaced by the Indo-European sky gods of the northern horse-based cultures whose gods dwelt in the sky rather than the earth. Their chief deity in the Hellenic culture was a male—Zeus.

She seemed genuinely interested in all this and knit her brow and fixed my eye, so I carried on. I went on to explain that even in the Hellenic culture of ancient Greece snakes continued to be revered and that it wasn't until the Judeo-Christian era that they became agents of evil.

"You know about Adam and Eve, right?" I asked.

"Of course," she said.

"Bad snake, right?

"Yes, gave the apple to Eve."

"Weak woman, Eve . . ."

"Yeah. So?"

So, I proposed, that in order to put to rest the ancient beliefs, the newer monotheistic Judeo-Christian religions had to demonize the earlier gods, including the powerful female snake goddess.

"Women became subservient," I said. "Snakes became evil." She stood silently for a few seconds, eyeing me. "Cool," she said. "But how do you know all this? Are you some kind of professor?"

I tried to explain, without sounding like a total nerd, that I had always been interested in nature and history.

In due time, after a little lighter conversation we went our separate ways, but I could imagine her telling her friends about this weird guy she met in the woods who tried to tell her that God was originally a snake.

I can't imagine what she would have said if I had told her one of the most bizarre of these various Christian transformations of pagan cultures. The third century prophet Manes used the traditions of local pagan serpent worshippers of Asia Minor to teach Christianity. He claimed that Christ was an incarnation of the Holy Snake who had encircled the Virgin Mary when she was a child.

· · ·

Down by the banks of the brook, I was getting uncomfortable observing my local water snake. Mosquitoes were beginning to bite; one landed on the back of my neck, and I reached around quickly and slapped it. The noise, or the motion, alerted the snake and it lifted its head, looked at me, and flicked its tongue. It apparently decided that I was too close and slowly uncoiled itself and slid along the bank into the grasses on the north side, where I saw it enter the water.

I had seen water snakes on my various summer canoe trips down the brook, and also ribbon snakes. As the resident snake woman had pointed out, these favor the stream banks and look very much like garter snakes, save that they are longer and narrower and tend to swim more than the garter snakes that also occur in the dry lands above the brook. The ribbon snakes feed on small frogs and tadpoles, newts, and small fish, and are skilled underwater swimmers.

I sometimes think of Kata herself as something of a snake goddess; at least she knows a lot about them. She picked up a lot of snake lore when she lived with the Hopi people. The most famous and, until the 1930s or so the most secretive of the Hopi rituals, was the famous snake dance, in which dancers shuffled through the village holding snakes, sometimes even carrying them in the mouths. They used a number of species of local snakes for the ritual, but the most extreme were rattlesnakes, which they managed to carry without being bitten.

Among the Hopi clans, the snake was considered a protector of springs and eventually, at least according to Western anthropologists, evolved into a rain god. The festival, which was held every other year in August, was considered a rain dance.

Kata doesn't have much respect for the modern Hopi snake dances, which nowadays seem to be performed mainly for tourists. She says she prefers the ancient ritual, which of course she has never seen but has heard about. The ritual was formerly secret even to some members of the tribe and was not seen by whites until the early twentieth century. In earlier times, she says, wild snakes were captured and "washed" with a snake whip, which was nothing more than a light feather with which snake priests would stroke their backs. The snakes were then kept in a large jar.

Around noon on the day of the dance, they were washed again, accompanied by rhythmic chanting by the snake handlers, and then the snakes were taken up by the dancers. Just before sundown, antelope priests would emerge into the central plaza and circle it four times; then, crouched over and stamping rhythmically to a slow steady drumbeat, the snake dancers would follow, holding the snakes in their hands and some carrying them in their mouths.

The modern dance was much diluted, Kata thinks, no longer secret and mysterious and probably no longer believed anymore by the people, who now rely on weather reports and don't really believe that honoring snakes will bring needed rain. Kata had learned about the ancient dances from the older residents, although most of them probably never saw the dances as they once were, having been born too late in the century.

There was an interesting river and water theory as to the origin of the Hopi snake dance put forth by the Yale architectural historian Vincent Scully, who believed that Native American dances were among the greatest of all the indigenous native art forms and were closely tied to the local environments. He believed that the writhing snakes of the dance replicated the winding rivers, streaming over the lowlands below the mesas.

Kata's attitude toward the modernization of this particular Hopi ritual, as well as certain Native American traditions, is not limited to popularization of the snake dance. She hates casinos and the fake Indian artifacts—most of which are manufactured in China—that are sold in tourist areas with a strong Native American presence. But then she dislikes most popular things of the modern world, including ball games, which were, as she accurately points out, an invention of the native people of the Americas. Now she sees sports as a convenient opiate of the people, not unlike religion, which she also questions. She cannot comprehend why millions of otherwise intelligent individuals collect together in vast temples, chanting and raising their arms in prayer as if to encourage the demi-god sport figures who attempt to hit a ball with a stick or, worse, smash into one another in order to get hold of a pigskin ball.

Kata is only one of a group of people in these parts who keep track of the ancient seasonal festivities, and the lore of older un-American, or non-American, festivals. These people do not make much of Thanksgiving, for example. Kata and her Wampanoag friends actually hold the equivalent of mourning ceremonies on Columbus Day. Many do not make much of the Fourth of July, and even less so Labor Day and Memorial Day. Nor do they seem to have much to do with traditional religions, although Kata told me she has a private god whom she worships whose name is Macombe.

It is hard to know sometimes when Kata is joking. She is a worldly person, well educated, and obviously intelligent, but she sometimes puts forth bizarre theories and explanations for things. Her god or goddess is apparently an all-encompassing nature spirit. She, Kata, sometimes lays wreaths or offerings at the base of a certain oak tree in Concord where she believes Macombe lives, and she says she has built temporary altars in places where she claims she has felt Macombe's presence. I have periodically come across little installations of this sort in my forest wanderings, small pyramids of stone, surrounded by wreathes of ferns and sacred evergreens, but Kata claims they are not her work.

As far as I can tell, Macombe is a little bit like the spirit of the Pygmies of the Ituri Forest, who have no overarching powerful male god but do respect the spirit of the plants and the animals they share the forest with. The closest deity, whom they do not exactly worship but who sometimes makes itself present, is something called the Molimo, which is nothing more than a growling voice that on certain nights circles the camps, snarling and chanting. The Molimo is just a local man making growling noises into a hollow bamboo log. Everybody knows it's someone from the band out there in the dark making the noise, but they practice a version of the willing suspension of disbelief and accept it as a forest spirit. This, incidentally, is how my friend Riggs explains belief in spirits. Traditional cultures, Riggs says, believe the masked figure used in various rituals is an actual god, even though they know it's just a local tribal member behind the false face. They accept the fact that the masked man or woman has been subsumed by the spirit.

Turtles are the other common family of reptiles that can be found along Beaver Brook. I've seen painted turtles, wood turtles, snappers, and the rare and endangered spotted turtle over the years. Also snapping turtles, and at the lower end of the stream the rare and endangered Blandings turtle. Painted turtles are by far the most common. Years ago, I made a turtle survey of this section of Beaver Brook. I put in at the Great Road bridge and paddled down to the second bridge, counting turtles all along the way, and counted over three hundred painted turtles in the three-mile stretch.

Along with snakes, Kata also identifies with turtles. She believes (perhaps?) that a giant turtle was the creator of the world and that we live on Turtle Island. It's true that, along with bears and snakes, turtles rank as among the most significant and common animals of religions and myth. The Eastern Woodland Indians believed that the Earth rests on the back of a huge primordial turtle.

Turtle Island, Kata says, is a local native term for the Earth.

According to the legend, one of the daughters of the Chi Manitou, or one of the Great Spirits who lived in the country above the sky, disgraced her father, so he cast his daughter through a hole in the sky and she fell into the wide empty ocean. She would have drowned, but a huge turtle rose up from the depths and she climbed onto its back. Nonetheless, she would have perished there without sustenance, but all the local water beasts took pity on her and decided to create land on the turtle's back. But how to do such a thing? They decided to dive down to the bottom of the ocean and bring up the sea floor. After much discussion, however, only one small animal dared to undertake the dangerous journey—the muskrat. He dove deeper and deeper and finally came to the bottom of the sea and carried solid earth up to the turtle shell. He made multiple trips and eventually plants and animals sprang up and the Earth was created.

Tom Doublet and his fellow Massachusetts Indians would have known of the legend, but Tom may have preferred the less complex story of a singular god who had somehow managed to create the sun and the moon and the stars over the course of a mere week. My advisors, Kata and Riggs, both claim that the only reason Tom Doublet and his people accepted Christianity is that it made life easier for them. Instead of propitiating all the animals who gave up their lives to hunters, or worrying about rain gods or rock spirits or tree spirits, they could just pray to one big god. "Doesn't mean they gave up on the old ones," Kata says. "They still believed in bear spirits and deer souls, but the new Great Spirit would care for them, they believed."

There is some contemporary evidence for this theory. The Mayan cultures of central Guatemala and parts of Mexico still maintain aspects of their ancient animistic believes, but they mix their rituals and ceremonies with their Catholic faith. The same traditions endure in Christianized Haiti and Cuba and the Christianized rural people of West Africa.

Locally, there was a contemporary local Native American woman who had renamed herself Tonapasquaw, or Turtle Woman, and one of the leaders of the Cape Cod Wampanoag tribe called himself Slow Turtle. So the legends must endure, even into our meretricious age of getting and spending.

The idea of Turtle Island is not confined to North American cultures. Pacific Islanders have a similar creation myth, and turtles figure in the legends and myths in most of the human populations in which they are found, especially in Asian cultures. In Chinese legends, the turtle was associated with the creation of the Earth. Along with the phoenix and the dragon it helped the giant god Pangu, who was born out of the cosmic egg, create the world. Also, the turtle was associated with fortitude, patience, and especially longevity, which of course are reflective of its behavior. It was, after all, the plodding steadfast turtle who, according to Aesop's Fables, won the foot race with the speedy hare.

But it's a dark and it's a heartless future we are facing without the presence of all these seemingly enduring species of the planet Earth. Even the long-lived turtle, whose heritage reaches back 260 million years seems to be in trouble, at least locally. Most of the snappers we see in this area are older— younger ones seem to be less common, and the spotted turtle is in danger, as is the Blanding's turtle, the local box turtle, and the wood turtle, which I have only seen once or twice around Beaver Brook. Thirty years ago in Connecticut where I used to live, all these turtles were common.

Global populations of frogs, which have been with us for nearly 300 million years, are also dying off because of a virulent fungus whose origins are obscure. Salamanders, too, seem to be in decline. I used to walk around the woods turning over rocks and logs to see what species of salamander might be sheltering there. But rarely do I find one anymore, and if I do, it would be the formerly very common red-backed salamander.

The greater Northeast is supposed to be getting more rain as a result of global climate change. Populations of reptiles and amphibians should be increasing as a result, but that does not seem to be the case.

Along with the drastic loss of species, worldwide plant and animal myths and legends have been diminished over the past two centuries, and one has to wonder whether future generations of children will live in a world where even the subjects of the ancient myths are animals of a distant past.

. . .

Late one afternoon toward the end of the month, I was standing on the Beaver Brook footbridge when I saw a small chunky dark bird rise up out of the green marshes and drop back down. The bird was unidentifiable to me, but I saw Riggs on my way home and told him about it and he said it sounded like a rail. I thought otherwise. I had never seen a rail on the Beaver Brook marshes, although I thought I once heard the call of a sora rail. The discussion of local birds carried on, and inasmuch as the day was almost over, Riggs invited me back to his porch for a drink. One drink turned to a second, and he began a long description of a nesting quetzal, sacred bird of the Maya, he had once found. This led to an account of an imperial eagle he had seen years ago in the Coto Donana south of Seville. Then he told me how he had also seen a lammergeyer while hiking in Pic du Midi, in the Pyrenees. And then, as he sometimes would do, he launched into one of his discourses, a litany of mystic and literary birds, starting with the seers of the Iliad predicting the outcome of battles by the formations of bird flocks and eagles, and the dove of Christian lore, the Ibis who was associated with wisdom in ancient Egypt, the eagle of any culture where eagles occur, and the raven who rides on the shoulder of Odin, and who was also the founder and creator of the Tinklit people, the peacock in Asian and Buddhist mythology, birds as messengers, birds as omens of a coming death, birds as pilots, and the ability of some birds to predict lightning strikes, and finally birds transforming themselves into gods. And then, after a reflective pause, he carried on with magical, imaginary birds, the winged Harpies of the Odyssey, the Phoenix, spirit of the sun, who consumes itself in fire and then is reborn from its ashes, and the eagle-like griffin, who has the body and tail of a lion and the head of a predatory bird, and also the firebird of Slavic lore, the massive roc of Arabic folklore, who fed on lions and elephants, and the giant thunderbird of Native American lore who cast lightning down upon the earth and the seas by flapping its massive wings. One more whiskey, and after—I confess—some encouragement from me, he began telling me animal tales of cultures he knew of, including helpful mice, sacred rats, lions, of course, and bears, rabbits, dogs, cats, bulls, and onto snakes, frogs, turtles, and toads.

By this time the sun was sinking below the forested ridge west of his house, and the eastern banks of the brook were glowing in the yellow light.

I finally extricated myself, and walked home in the twilight, the evening song of robins and mocking birds echoing out of the woods beside the road.

CHAPTER NINE

THE THUNDER MOON

Henry Thoreau, for whatever reason, spent a lot of time in his boat on the three rivers of Concord in August of 1858. Almost every day, it seems, he was out on one river or another, studying the progress of the plant life, and recording the activities of birds and fish. At one point he heard the regular pounding call of a bittern, which in his time was known as the stake driver, and he saw great blue herons, and a snapping turtle, and he observed the habits of a mud turtle or stink pot, which actually climbs up onto the overhanging branches of riverside trees. They would tumble off their posts whenever he approached with his boat.

Then as now, the water lilies were in bloom, and as usual Henry observed them closely, so closely as to determine that the sweet odor of the lilies that flowered in Concord were richer than the scent of the water lilies in Tamworth, New Hampshire, where he had also seen them that year. He was, for all his poetic mysticism and literary interests, an astute and exacting naturalist. He measured the physical structure of different flowers, remarked on the length of bloom of various plants, and also, and always, included the dates of their blossoming.

He also records, as was his custom, a little cynical aside on the Follies of Man. He would regularly see what he termed a pair of "summer ducks" out on the river that year—probably mallards. These were fairly tamed and he was able to get close to them in his boat, but then he stopped seeing them. It turns out a certain Mr. Goodwin had shot them and, as Henry records in the journal, his wife ate them. She (presumably) enjoyed the flavor of the duck, whereas he, Henry, enjoyed their beauty. Thus, he writes, they both in their own way were "fed."

The significance of this metaphor is left unsaid.

Beaver Brook itself is in full fecundity at this time of year. These are the so-called dog days of August. The heat is on, and the fruits of all the flowerings and matings of June and July are everywhere, settling over the land and marshes. The fertility of the brook is such at this season that the very air is rank with the smell of living and decaying vegetation and rising swarms of midges and dancing flies. In the quiet coves, the duck weed spreads over the surface of the waters, and the arrowheads, blue flags, and the water lilies are in bloom, and the swifts and swallows are coursing above the marshes preparing themselves for the long migratory flights ahead. Turtles are everywhere, too: huge, Jurassic snappers, painted turtles, spotted turtles, and the occasional musk turtle. Herons, bitterns, and snipe are still around, and the garter and ribbon snakes and the brown snake lay their eggs at this time of year. Turtlehead blooms and the heavy scent of the flowers of clethera fill the air. Swamp mallow flowers at the edges of the cattail stands and the thickets of sedges and rushes and reeds. Leopard frogs and pickerel frogs disperse. Red-winged blackbirds and marsh wrens and swamp sparrows rise and fall from the thickets of cattails and button bushes, and along the edges of the marsh spiderlings hatch and begin to pay out their gossamer silken threads.

Early August marks the ancient festival of Lugnhasah among Celtic cultures, the midpoint between the summer solstice and the autumn equinox. It was originally celebrated with feasting and fairs and, as always during these solar dates, bonfires, and also, as usual, dancing, in this case circle dances, and in England the reenactment and ballads of John Barleycorn, whose "head" was chopped off to be made into whiskey and beer. Folklorists have placed him in the company of those ancient vegetation gods and goddesses who suffer, are killed, and then are resurrected.

Lugh was the ancient King of the Sun, and there was a sense in this festival that he was slowly losing his power. It was a time of change. The grain was ripe, and the swallow flocks were collecting, and darker days lay ahead.

August is also the month when Sirius, the Dog Star, rises, marking the beginning of the dog days of summer, the hottest time of the year. Although hardly recognized in our time, this was a major event for ancient cultures; it was associated in Egypt with fertility and the goddess Isis, and in Greece and Rome with celebrations associated with various dog-loving gods and goddesses, including the powerful goddess Hekate, who, among other things, helped guide Ceres through the Underworld in search of her daughter Persephone.

Sirius is the brightest star in the constellation Canis Major; fittingly, Canis, the Dog, arrives in the sky ahead of the winter constellation Orion, the hunter, as do all good hunting dogs.

On one of those hot August evenings, I took my young son out in the boat at the lower end of the brook to try to introduce him (in case he cared) to fishing. I personally had never been interested in catching fish. I always used to feel sorry for the ones I did manage to catch, and the thought of a hook in your lip was horrifying to me. Nevertheless, I thought I should try to teach him American values, one of which, I suppose, is fishing.

We paddled upstream, amid the insect swarms and the darting swifts and tree swallows, and he sat in the bow, casting ahead and not catching any fish, even though we knew they were down there somewhere in the murky depths. Turtles were still out, and the marsh wrens and the red-wings were everywhere, screaking and twittering, and the swifts were dashing and diving all around us.

At one point my son made a high cast and, by chance, reeled in not a fish but a swift, which must have made a dive for what it thought to be a large insect. My son reeled in the struggling bird across the water, and I caught it up and untangled it. I held it in my hand for a bit, its heart hammering against my curled fingers while we studied it. And all the while the bird fixed us with a cold but seemingly intelligent eye. Then I raised my arm and opened my palm and it spirited off across the grassy marsh twisting left and right as it fled.

We quit fishing after that but carried on upstream in the settling light, enjoying the evening and the lush, grassy tunnels of high cattails through which we were paddling, and then we turned around and floated slowly back toward the green evening sky. Swifts, tree swallows, and, by that time, bats were everywhere above us in the fading light—living things of the air, streaking here and there, and then dashing low and twisting upward, ever twirling and diving, and rising again. One could understand at such times why the people of ancient cultures around the world believed that birds and bats were the souls of the dead.

Where we go when we die is one of the great questions of the human condition that rose in the dim beginnings of human consciousness. Even the supposedly dull-witted Neanderthals struggled with the idea. They may have placed rings of plants and fungi around the bodies of the deceased and decorated their gravesites with the bones of sacrificed animals as well as weapons and tools. They must have understood mortality and believed that the visible world of everyday life was not the only reality.

In what is believed to be the most ancient Cro-Magnon ceremonial burial site, the skeleton of a twelve-year-old boy was surrounded by chipped bones, and in another 28,000-year-old grave site in Sungir, Russia, archeologists discovered evidence of the ritual burial of an adult and two children, offering evidence of religious practices. Burial artifacts of this sort reflect the formerly close association of early humans with their natural world. Flowers, bear skulls, and later dogs, mummified cats, and, in Scythian cultures, even horses have been discovered in ancient human grave sites throughout the world. Often the remains of food stuffs are found among these artifacts, providing sustenance, it is believed, for the long dark voyage of the afterlife.

Later, more advanced (supposedly) cultures describe a specific place where the dead would go, an Underworld of living shades, or a Heaven in the sky above, with angels, harps, and clouds, or a Paradise, with fountains flowing streams and virgin maidens, reincarnation, and then Nirvana, and the final release from the round of death and rebirth, and also Valhalla and the halls of Odin, or, in some religions, a dark Hell where one is tortured for eternity, and finally, according to the religion of science—nothing.

Death is an event horizon; no one knows what lies beyond.

. . .

My general custom in summer months is to get down to the brook early in the morning, if I can. But in June and early July, this is not always easy. The sun comes up over the eastern bank of the brook at five in the morning, too early—especially after a late night—to get down to the brook before sunrise. But by August, sunrise is a little more reasonable. I would go down to the brook after coffee but before breakfast, hang out for a while, and then return to the garden for more coffee and a late breakfast.

There is a tea house at the southwestern end of the garden, and I often retire there later in the morning and listen to sounds of the woods and fields. I can hear traffic out on the Great Road if the wind is right, plus passing cars on the road below the house. But I can also hear, and sometimes see, the domesticated animals of the Scratch Flat farms, not including my own dog and cats. The neighboring dogs often sound off periodically. I can hear my neighbor's roosters, and also—when the wind is right—the whinny of a local horse. Over the hill, closer to the farms on the western slopes of the ridge, I can also sometimes hear the lowing of cattle, and the calls of sheep and goats.

These are, to say the least, ancient, familiar sounds, the natural background noise of rural communities the world around, and as with so many other plants and animals, these common, domesticated animals have also entered into the cultural myths of the world.

Dogs, and slightly later, cats, are generally considered to be the first domesticated animals, although recent studies of ancient steppe cultures place the domestication of the horse in the same general time period. All these animals, followed by sheep and goats and chickens, ducks, and geese begin to appear about 7,000 years ago in the archeological records.

Dogs are still generally considered to be first. Why and how this happened is still under debate. The general theory, now somewhat in question, is that either jackals or wolves began following human hunters, having determined that at some point the hunters would make a kill and there would be leftovers after butchering. In time, the wolves or jackals began to assist in the hunt by following the scent of prey and would then follow the hunters back to their camp, hoping for more. The atavistic behavior of modern dogs offers evidence of this. Anyone who goes out for a local hike through the woods with an unleashed dog may have noticed that on the way out the dog will go first, sniffing and searching. But on the return trip, it will stay with its owner. The point being that in ancient hunting cultures, the dog helped find the prey and then followed the hunters back to the camp for the scraps.

Cats, it is assumed, became domesticated when early agriculturalist communities began storing grain. The grain attracted mice and rats, and so the local wild species of small felines began to collect around the storage places. Eventually they tamed themselves and became accustomed to human beings.

Ritual burials of cats and horses turn up in archeological digs as far back as 4,000 BCE; dog burials occurred even earlier, although these may have been pets or favored dogs rather than ritualistic burials. Formal burials of cats and horses suggest that these animals were honored and even figured into myths and religions. Archeological evidence from Egypt indicates that the cat was held sacred as far back as the First Dynasty, roughly 4,000 years ago. The cat-headed god, Bastet, originally a lioness figure, was transformed into a domestic cat, and cats were so revered that they were mummified and buried along with noble figures.

In spite of their deep relationship with human cultures, dogs do not always fare well in mythology and folktale. The vicious three-headed dog, Cerberus, guards the entrance to the Underworld in Greek mythology, and a variety mythic dogs, in a variety of cultures, always seem to play a role as guardians of wild places, the hunt, and the Underworld. In particular, in Northern Europe they were often associated with the so-called wild hunt in which, on certain windy nights, ghostly bands of hunters, led by baying hounds, streamed over crossroads and ravines and hills, sweeping up anyone caught in their path. In some cases, these unfortunates were dropped back down on earth when the wind died and would have no recollection of where they were or what had happened to them. The hunt appears in Norse mythology in Odin's Hunt, in which hunters with horses and barking hounds ride the upper airs carrying the souls of the dead to the afterworld.

The wild hunt was also a standard in Welsh folklore. Ghostly hounds known as Cwn Annwn, the dogs of Annwwn, ruler of the Underworld, coursed over the land during the windy days and nights of the twelve nights of Christmas. Those who heard the baying of the hounds were bound to experience a death in the family.

A variation of the wild hunt is even recorded in the early days of colonial settlement in North America. In Plymouth, just before the outbreak of King Philip's War in 1675, there were all manner of ominous warnings, including ghostly troops of horse companies coursing overhead in the night sky.

A black doglike beast known as the Hellhound occurs in a variety of northern European folktales; it is an ugly, black dog with matted fur and glowing red eyes and it haunts graveyards and entrances to the Underworld. It is a phantom being and lives between the spirit world and the everyday world of human beings. And, of course, it's dangerous. If you look into its red eyes, you will die.

Generally, dogs do not appear to be favored in the folk-lore of the British Isles, and black dogs seem to head the list. The tales vary from town to town and in different regions, but despite its different names, its basic form is the same. It is a mysterious gigantic black dog, sometimes with gleaming red eyes, and when it appears someone is certain to die. Conan Doyle used a local legend from Devonshire to set the stage for *The Hound of the Baskerville,* and if his description of the hound is in any way reflective of the legends, which may still have been abroad in Conan Doyle's time, the beast was truly hideous: "A great, black blazing-eyed beast, shaped like a hound, yet larger than any hound that ever mortal eye had rested upon . . ."

In the novel, the hound is known to haunt the moors beyond the Baskerville estate, and one night after a drunken revel, the reprobate lord of the manor rides off in a drunken escapade to catch up with a virtuous maiden who has escaped from his clutches. A terrified shepherd happens to have seen him pass and recounts the fact that he was pursued by a large, frothing, dark, doglike figure the size of a small horse. Later, the lord's fellow revelers catch up and find him. The Hound has killed the lord and stands above him tearing out his throat, blood and froth dripping from its jaws, eyes blazing.

The dog legends seem to have endured into modern times. I once spent some time on a farm in the Yorkshire Dales, and early one evening I walked across one of the dales to a nearby pub to lift a glass. There I met a bushy-haired old man and we fell into conversation about the historical land division system in the Dales and other local lore. It was a dark moonless night by the time I left to walk home.

"You'll be watching your back on the road home, boy," he said.

I asked him why: Were there highwaymen about?—which seemed unlikely in that rural and peaceful region.

He lowered his head and fixed my eye over his glasses. "It's the hound," he said. "The Barghest Hound is about on nights like this."

The Hound, it turned out, was a gigantic black dog with long fangs that haunted the moors. It usually fed on sheep and even cows but was known to attack solitary walkers from time to time. I'm not sure whether my companion was joking or actually believed what he was telling me. But either way, I got back safely to the farm where I was staying.

Not all dogs of myth and legend are bad, though. Odysseus had a loyal dog, Argos, who, though old and sick, recognized him when he returned home after his long ordeal. And Plato argues that the dog is the most philosophical of all the animals. It has a system of beliefs. If it knows you, you are good. If it doesn't, you bear watching.

Dog-headed gods appear in glyphs on the ancient walls of the ruined pyramids in the Nubian city of Meroe, and later the Egyptians honored the dog-headed deity, Anubis, who was the god of the Underworld and protector of the dead. He later was associated with the embalming process. Anubis may have originally been a jackal, but the domesticated Egyptian hound looks very much like the wild jackal, with a long nose and pointed ears.

Dogs also fare well in Hindu lore. The dog god Shvan was an honored figure in one of the many Hindu festivals, and it is believed that kindness to dogs will help you get into heaven. The belief probably evolved from a famous scene in the Hindu classic the Mahabharata, the ancient Sanskrit epic of a long war between cousins. In the tale, at the end of their lives the five Panadava cousins begin their final pilgrimage up through the Himalayas on their way to heaven along with a dog, the companion of King Yudhisthira. At the end of the journey, a blazing chariot appears before the king and Indra invites Yudhisthira to enter heaven. The king and his dog mount the chariot to enter, but Indra resists. "Leave the dog behind," he commands. "Dogs cannot enter heaven."

The king dares to argue. "How can I abandon this loyal companion who has followed me all this way?" he asks.

A discussion begins on the nature of earthly attachments when suddenly there is a violent flash of lightening and the dog reveals itself as the god Yama. He praises Yudhisthira for his loyalty and welcomes him into heaven.

There was also a benevolent cat god named Shasti in the folklore of India. She was the protector of children and was also associated with vegetation and the harvest. And there is a singular and famous wooden cat statue in Native American culture, an artifact of the pre-Columbian Calusa people of south Florida. Coincidentally, the statue looks very much like the many carvings and statuettes of ancient Egyptian and Mesopotamian cats.

After the decline of the ancient Egyptian cultures, cats do not fare well, at least not in Northern Europe. In the seasonal festive fires of pre-Christian Europe, cats were thrown into the bonfires to assure the future health of sheep flocks. Later, the Church determined that cats, especially black cats, were agents of the Devil, and so they continued to be burnt in the holiday fires, often in the cruelest way possible. Snakes, the other associates of the Devil, were also rounded up and thrown into the fires.

But of all the domesticated animals of the ancient world, without compare, the horse was the most important. Dogs have their place as hunters, and cats kept mice and rats at bay, and later in history both these animals evolved into companions of modern human beings and recently have proved to offer solace and even a health benefit to people. But the horse changed history.

The general consensus holds that the horse was first domesticated in the steppe region above the Black Sea in Ukraine and southern Russia by the horse-hunting cultures such as the Sredni Stog, which lived in the region about 6,000 years ago. Burial sites of these Central Asian cultures reveal horse skeletons, and archeologists studying the wear marks on the teeth of horse jaws determined that the horses were ridden by the fact that the back teeth were worn down by the bit.

This singular advancement—the ability to ride horseback and cover wide distances—opened the world to the riders of the steppes. It is even credited with the spread of Proto-Indo-European languages. Warriors on horseback could now range out of the seemingly endless grasslands, conquer distant tribes, and spread their culture as well as their language.

The language of these ancient steppe cultures was the distant ancestor of all the Indo-European tongues, including Sanskrit, Greek, Latin, and all the modern languages of Europe save for Finnish, Basque, and Hungarian.

Given the importance of the horse in early Bronze Age history, it is not surprising that horses figure in the mythologies of cultures around the world. Even before they were domesticated, the wild horse was honored with highly realistic paintings on the cave walls of France and Spain, and after domestication they were so ingrained in the steppe cultures that they were buried in ritual fashion with grave goods, often accompanied by two dogs. These sacred burial sites were marked by a slanted tripod mounted with a horse skin and head, some of which were still being constructed by local tribes even into the twentieth century. The burial sites and the presence of dog skeletons conform to an ancient European belief that a horse carried the souls of the dead to the gates of the Underworld, which was traditionally guarded by two dogs.

Horses figure in the sacred and profane mythologies of any of the cultures in which the horse was used, and its importance was recognized in images and statuettes reaching back thousands of years. The famous White Horse of Uffington, in England, which was created by scraped turf revealing the white chalk substrate, is over 3,000 years old. Traditionally white horses were favored, some associated with the sun, some with fertility.

Most famous of these, at least here in the West, was the Greek winged horse Pegasus, the stallion child of the hideous Gorgon Medusa and the sea god Poseidon. In fact, Poseidon was the creator of all horses; he fashioned them from the cresting waves of the ocean.

Pegasus was caught and tamed by the rider Bellerpheron and accompanied and even assisted him in many adventures, including a fight with the monster, Chimera, who was a combination of a fierce lion and a goat and, dragon-like, could spew out fiery breath. At some distant point, Pegasus struck the earth on Mount Helicon and left behind a horse hoof–shaped spring known as the Horse Well, from which flowed the sacred waters of Hippocrene. The spring was favored by the Muses, and poets would drink the waters for inspiration.

A white horse also figures in the Hindu pantheon and was associated with the sun god Surya. The god Indra favored white horses, and pale horses appear in the myths and legends of many other Indo-European cultures. But honored though they were, these same cultural groups were not averse to ritually sacrificing horses, especially in the Proto-Indo-European steppe cultures where the horse was first domesticated.

Kata believes she may have seen one of these mythic spirit horses in the Estabrook Woods, in Concord. She was out for a walk in the woods early one misty autumn morning when she heard the thunder of hooves. A large wild-eyed white horse, accompanied by two dogs, burst out of the fog and swept past her and disappeared into the mist without breaking stride.

"They came and went so fast," she said. "I couldn't be sure they were real, save that I heard the hoof beats."

• • •

Henry Thoreau may have been familiar with some of these tales. He was versed in Hindu literature and cites the Vedas in *Walden*, and, like most of his educated compatriots, he could read ancient Greek and Latin and was familiar with the Greek myths: they permeate some of his more florid published descriptions. He saw Mount Olympus, home of the gods, in the minor hills of western New England, and when he reached the heights of Kahtahdin he imagined that he had accidentally entered into the wilds of the Caucasus Mountains, where Atlas was chained.

He was especially fond of cats, and also dogs, and had a way, it is reported, with wild animals. He had a tame mouse that would visit him in his house at Walden Pond, and he could call forth woodchucks and squirrels with a whistle, and was often visited by a pair of crows who would sit on his shoulder and feed on crumbs. The Concord locals began to gossip about him as some sort of Orpheus, who could tame wild animals with his music, or a modern Saint Francis, who went so far as to preach to birds, and famously tamed a mother wolf.

Henry was also alert to all the voices of the local fields and forests. At one point on an August afternoon, he reports hearing somewhere below his feet the otherwise obscure and ordinary sustained trill of a mole cricket. Late August in Henry's time, as it is now, was a season of insect songs. On summer days on the grassy landing I too could hear the calls of meadow crickets and mole crickets, and as dusk fell I could hear the singular chirp of field cricket, the shushing calls of katydids, and the steady, sultry throb of snowy tree crickets. The sounds of the mole cricket, along with the sustained trill of the meadow crickets, are the first clarion calls of the onset of the great chorus of stridulating insects that begins around the middle of the month. By late August the chorus is in full voice.

Down by the stream bank at night you can hear the ringing pouring out from the forested banks: katydids in the oak trees, bush katydids in the shrubbery, field crickets—and also the steady heartbeat pulsing of the snowy tree crickets. They produce a steady throbbing dull whistle-like chirp that responds, as do all the rhythmic calls of this group of insects, to the local temperature—the hotter it is, the faster the beats. According to Riggs, Nathaniel Hawthorne wrote that if moonlight could be heard, it would sound like the snowy tree cricket.

But the onset of this chorus marks the beginning of the end. Down by the brook one night in the middle of the month, I saw a shooting star dart across the eastern sky, then another, and then shortly thereafter yet another. It struck me that this must be the fifteenth day of the month, the time roughly when the Perseid meteor showers fall.

The barn swallows were gathering; southbound nighthawks were coursing through sky at dusk, the Joe Pye weed was blooming, the scent of clethera filled the air on the stream banks, and it was clear that summer was winding down.

CHAPTER TEN

THE HARVEST MOON

I took a night walk to the bridge one evening early in September and lingered for a while on the bridge, watching the waters slide along through the grass-wall tunnels of cattails. It was hot that night and the stridulating insects were in full voice on the wooded bank and in the old field leading to the bridge. It was clear and moonless, and at one point, staring down into the sheer black waters of the stream, I saw a faint twinkle, not unlike the clouds of bioluminescent jellyfish you see in coastal waters at this time of year. The mirage reminded me of a similar late summer night years earlier when I was making a boat delivery with my older brother, who, among other eccentricities, used to climb up to the crosstrees of the mainmast and execute swan dives. We were anchored in a quiet cove one hot night in the Elizabeth Islands, and he decided to go for a swim. He climbed the mast, stood out on the crosstrees, paused, raised his arms, and threw himself out into the black night and down into the dark waters. In the same seconds of his dive, a shower of Perseid meteors happened to fall behind him, and when he hit the water his whole form was trailed into the depths by a wide wake of glittering plankton. It was as if some sky god, or some fallen angel, had chosen to dive from the night sky down into the sea, trailed by stars. I later wrote an essay about the event that I titled, fittingly, "The Apotheosis of My Brother Jim."

This was no salty cove I was staring into, though; it was Beaver Brook, a fresh water stream, and it took me a while to realize that the flashes of light in the waters were the reflection of stars.

The brook makes a wide eastward bend upstream from the bridge before turning westward, leaving an open view of the southern sky above, and there overhead I saw the constellation, The Swan, her head pointed due south at this time of year, like so many earthly migratory birds.

According to the ancient Greeks, the constellation was associated with the myth of Leda and the Swan, the subject of many, sometimes controversial, Renaissance paintings. In one of his many animal guises, Zeus transformed himself into a swan in order to seduce Leda, the Queen of Sparta and the mother, in some versions of the story, of Helen of Troy.

The Greek swan god in the form of Zeus is just one in the body of legends dealing with swans. In both Hindu and Celtic lore, the swan embodies the connection between the unseen spiritual realm and the profane world of everyday life, which in some ways, I suppose, is true also of Zeus and Leda. Swans are usually benevolent and considered beautiful, as with the swan maidens of German mythology. The traditional swan maidens were actually women or girls, who for various reasons—usually as a result of a curse or a punishment—were condemned to live as swans by day and women at night. The beautiful swan child Brunhilde, of Wagnerian opera fame, was a classic swan maiden, as was Odette, the swan maiden heroine of *Swan Lake.* She was cursed by the sorcerer Von Rothbart and had to live as a swan by day and transform into a woman at night. But she and Prince Siegfried, who was at the time betrothed to someone else, fell in love, and as is so often the case in romantic ballets of this sort, things went downhill after that, with tragic results. They both die.

Folkloric swans are not always helpless and benevolent, however. Russian pagan religion has a swan maiden of death in its pantheon. She appears in the epic *Prince Igor* sweeping down over the Russian warriors spreading discord and death. And the swan in the Finnish epic of the Kalevala is hardly sympathetic. In one part of the cycle, the young hero Lemminki attempts to kill an evil black swan who lurks in the waters of Tuonela, the river of the Underworld. Lemminki drowns in the process and is lost in the depths. But as in other lost child folkloric motifs in both Greek and Egyptian mythology, his mother searches the world to try to find him. She eventually locates him at the bottom of the river, a broken man. Then, as with Osiris, she dredges him up and sews him back together.

The nineteenth-century Finnish composer Jean Sibelius wrote a tone poem called *The Swan of Tuonela* that, appropriately, is filled with dark, minor phrases. The swan is also used in Camille Saint-Saëns's *Carnival of the Animals* in a haunting duo for cello and piano. The melody was used in a solo ballet performance called *The Dying Swan,* most famously performed by the Russian ballerina Anna Pavlova.

In Romantic folklore, swans sing only once in their lives, just before they die, giving rise to the expression "swan song." How the legend developed is not clear given that even the so-called mute swan makes a guttural trumpet-like call, and the fact that they don't sing at death has been known since Roman times. But, as with so much folklore, it makes a good story.

There are other evil black swans in European mythologies. Odile, the black swan of *Swan Lake* and the daughter of Von Rothbart, was transformed by her evil sorcerer father to appear to be Odette in order to trick Prince Siegfried away from the real Odette. The prince and Odette are only liberated by death in a double suicide, after which they (of course) ascend to the heavens.

This dark side of swans, at least here in North America, may be a more accurate characterization of the bird. The mute swan, which is now the most common swan species in the region, is an introduced alien and can wreak havoc with freshwater ecosystems since when it feeds it tends to uproot the aquatic vegetation. They are also notoriously aggressive and will attack human intruders who wander into their nesting area.

A pair of swans used to nest in the reeds near the open waters of the little bay on the downstream side of Beaver Brook bridge and would periodically feed upstream with their young on the eastern side of the bridge. I sometimes launch my boat at that point and paddle upstream, and in mid-summer I have to watch out for them. Once or twice one of the parents eyed me suspiciously from the marshes and started toward me, its wings held aloft, but I managed to paddle hard around one of the bends—out of sight, and presumably out of mind, such as it may be for swans.

Swan maidens and their ilk are not the only sometimes dangerous water beings. The Naiads of Greek mythology live in streams and pools and often charm young men and then drag them down into the depths of the waters.

Hercules, one of the members of Jason's Argonauts, had a serving boy named Hylas whom he loved dearly. Everybody, it seems, appreciated Hylas; he was a beautiful boy with golden curls and deep blue eyes. One afternoon, during the quest for the Golden Fleece, Hylas wandered off and came upon a spring called Pegae that was haunted by Naiads. When he got to the spring, he kneeled and bent over to drink. The water nymphs had seen Hylas as he approached and they too were smitten, and when he bent down to drink, they grabbed him and drew him down into the cold, glimmering cave where they lived. Heracles was devastated when he did not return and went out calling his name, but Hylas, trapped deep in the watery cave, could not hear him.

Whoever decided to find a generic name for the American species of frog known as the spring peeper must have heard in the plaintive whistles of the frog the cries of Hylas and named it *Hyla cruciform*.

Years ago, I had a girlfriend who claimed she was part Naiad. There was a clean river not far from where we were living at the time and we used to hike up to an isolated cove, overgrown in sections with water lilies. She used to swim naked out to the patch, dive down into the murky green, rise up draped with lily pads and white flowers, and tempt me to join her. I would, of course, always refuse—knowing my fate if I did.

For all I knew, she might also have been a snake goddess. She used to try to catch any snake we happened to come across on our woods walks. I was swimming with her one day in another murky pond, overgrown with watery vegetation. At one point a black water snake slid down from the shore and headed for her, directly. Rather than flee or splash the snake away, as would any a traditional American girl, she held out her arm. The snake approached, slithered along her arm, crossed over her shoulder, which was more or less at water level, and disappeared into the depths. She later moved to Georgia, where I heard she was somehow involved in nature education and where there are more species of snakes than in Connecticut.

Naiads are not the only spirits that live underwater in streams and pools. The Undines were also beautiful water creatures that had the ability to take on human forms. Some of them longed to be human, however, and obtain a human soul, which they could do by marrying a faithful human man. They would leave the waters and assume human form to find a husband. But if ever the husband was disloyal to the Undine, he would die.

Nixies were similar creatures, common to Germanic cultures. They lived in vast underwater palaces, but would leave their watery homes sometimes and appear on earth in the form of beautiful women, or, in some cases, old weathered crones. In either form, they had the ability to prophesy. But they too could be dangerous. Like certain malevolent woodland faeries, they would sometimes steal children and take them down to their underwater castles. They were also said to sometimes marry humans and bear human children.

Many of these freshwater creatures are similar to the seagoing mermaids and are also similar to the Selkies, or the seal women of Scottish folklore. They too would come ashore, shed their seal skin, and appear as beautiful women who would marry good men and give birth to human children. But in order to hide their origins, they would hide their shed skins from their husbands. In a common tale in Scottish lore, the husband finds the skin, often in a trunk or a drawer that he was told never to open. If he found the skin, the Selkie would have to go back to the sea, taking all her children with her.

Nasty though some of these watery spirits could be, they cannot compare with Grendel and his mother in the English epic *Beowulf.* They live in a cave at the bottom of a deep lake surrounded by cliffs, where demons and fiery worms and all manner of devils lived. Grendel has been raiding the mead hall of the king of the Danes, carrying off warriors and devouring them. He is said to be the offspring of Cain and he hates humanity. The hero, Beowulf—whose name, incidentally, translates as "Bear" or "Beewolf" in Old English—arrives, and in a hand-to-hand fight manages to tear off Grendel's arm. The beast man flees to his underwater cave, where he dies. Enraged by the murder, Grendel's mother herself begins raiding the hall. But Beowulf, carrying his powerful magic sword, swims down to the cavern where she lives and encounters the demon mother. An epic battle ensues in which his magic sword fails and Beowulf is nearly done in, but he snatches the sword of a giant from the cave walls and manages to decapitate the hideous, frothing monster.

And so begins early English literature, with its wealth of dragons, giants, witches, wizards, faerie queens, demon lovers, questing knights, cursed castles, water monsters, the Green Man, and the Green Knight, and perhaps the most enduring figure of them all, Robin Hood, who is in some ways emblematic of them all—a shape-shifter, trickster, forest-dwelling guardian of Sherwood Forest.

• • •

In the same bend on Beaver Brook, just north of the bridge, there is another mythic bird that often perches on a tree branch above the brook, the kingfisher, the benevolent Halcyon of Greek mythology.

According to the story, the god of winds, Aeolis, had a daughter named Halcyon, who was deeply in love with her husband, a mortal named Ceyx. Things never go well with happy families in Greek mythology, though. The couple was deeply in love and used to joke around and call each other Hera and Zeus. Not a good idea, verging perhaps on the equivalent of blasphemy, so Zeus caused Ceyx to drown in a storm at sea. Halcyon learned of his fate, ran to the sea, and found his body. Disconsolate, she threw herself in the ocean and she too drowned.

But the gods took pity on the couple and turned them into kingfishers. The only hitch was that Zeus commanded that the couple nest in winter. But Halcyon's father, the god of winds Aeolis, promised to keep the seas calm during their nesting season—thus the origin of the term "halcyon days."

According to Riggs, kingfishers also make it into Arthurian legends of the Grail quest, which was presumably drawn from earlier Celtic and German lore. The earliest written account of the tale is from Chrétien de Troy's twelfth-century poem *The Story of the San Graal*, in which, on his way to visit his mother, Sir Perceval meets the Fisher King. The king invites Perceval to stay at his castle, which turns out to be an enchanted domicile in which, before each meal, a procession of young men and women pass though the dining hall carrying apparently sacred objects. The last of these young people is a beautiful maiden who carries an elaborately festooned cup, or grail.

After his visit, Perceval resumes his travels but soon meets a dejected young woman who, somewhat mysteriously, suggests that he should have asked more questions about the Grail. He is mystified by her statement but later meets a hermit who explains the mysteries. The cup is in fact the Holy Grail, which has miraculous curative powers, and if Perceval had asked the appropriate questions he could have cured the Fisher King, who has been languishing for years with an unhealed wound in his groin.

After Chrétien de Troy, the story of the Grail Quest became an important literary theme, adapted and more Christianized in the Arthurian story cycle of the Knights of the Round Table and used in poems and *chansons des geste* from the tales of Sir Galahad all the way up to T. S. Eliot's epic poem *The Wasteland*.

Why the old king is wounded and remained so is not known, but his wound, which seems to have made him impotent, is extended to his whole kingdom; the land lies infertile, a veritable wasteland.

Riggs says the kingfisher enters the tale in some of the earliest versions in the form of a word play in the medieval Romance languages. In French, the Fisher King, *Le Roi Pêcheur* (with a circumflex), is easily transliterated as *Le Roi Pécheur* with an acute accent, meaning "The Sinner King," which may help explain the incurable wound.

The English name for the bird evolved from the "king's fisher," and legends, other than that of the halcyon days, are many. It was once thought to be a dull gray bird but, according to medieval legend, after liberation from Noah's Ark, it flew toward the sun and acquired its blue back from the sky and its reddish underbelly from the setting sun. Its dried body had the ability to avert thunderbolts, and if it was hung from the ceiling it would point its bill into the wind, even though it was sheltered indoors.

In autumn, kingfishers leave northern regions when ice forms; I sometimes still see them on the lower reaches of the brook in late October. Some move to the coast, but eventually they fly southward as the ice begins to set.

• • •

On the east side of the brook, downstream from the landing, the stream narrows and runs hard between two large boulders on each side of the brook. It was here, probably, that Tom Doublet maintained his father's fish weir, and sometime in the late eighteenth century, it is likely that a family named Frost created a ford to get over to the western side of the brook at this point.

Although it's not easy, you can haul a kayak or canoe out of the brook at that spot and bushwhack up though the tangle of brush to a stone wall that runs in a generally east-west line.

New Age sojourners in this area claim that this wall, and others not far from the site of the Nashobah Christian village, were laid out by the Native American people in alignment with the setting sun on the twenty-first of September, the day of the autumnal equinox.

Academic archeologists question the idea. Once the forests were cleared in the late eighteenth century, the local farmers tended to construct walls everywhere, some to keep sheep and cattle in, some just to pile the many rocks they were forced to clear from agricultural fields. Inevitably, some of these walls would line up with the solstices or the equinox in spring and autumn, according to archeologists.

Furthermore, for more than 8,000 years of Native American occupation, the land on either side of the open floodplain of the brook was thickly wooded with oak, maple, hickory, and chestnut, with an occasional lone white pine spearing above the hardwood canopy, so it's unlikely that the sun was even clearly visible at the time of the autumnal equinox when the leaves were still on the trees.

That is not true, however, in other parts of the world, where either because of agricultural conditions or open natural prairie or deserts, the setting sun was visible. And here, in these cultures, the autumnal equinox, which often occurred at the time of the harvest, was a time of celebrations and harvest festivals.

In the British Isles and elsewhere throughout Europe, this was the origin—or the excuse, at least—for harvest festivals, which were associated with many legends and traditions, including the European belief in a mythical beast known as the corn wolf, a creature found in the last sheaves of the cut grain or the last of the bundles to be stored. The wolf did no immediate harm but was associated with bad luck. Women, who were often the harvesters and the binders of bales, were the most affected.

In the Welsh tradition, the autumnal equinox was known as Mabon, the opposite of the Celtic Ostara, the spring equinox. Mabon was associated with the onset of darkness and the shortening days that would last until the return of the sun on the twenty-first of December. The Druids held ceremonial processions at this time of year, and throughout Europe the ancient figure known as the Green Man made his appearance in association with the harvest.

The Green Man had his beginnings, it is believed, in the Neolithic period and was originally a consort or associate of the mother goddess. He reappears in Roman myth and by the time of the Middle Ages was an integral player in peasant folklore. His image, a man dressed in ivy and with a circlet of green leaves, appears even in the sculptural carvings in Christian churches, a holdover from the old pagan traditions. He is generally considered a benign figure, a man who lived between the world of the wild forests and the agricultural lands. He would sometimes lead lost children home, and he appeared often at harvest time.

He was in essence a primordial wild spirit of undeveloped forest. Enkidu of *The Epic of Gilgamesh* was a similar figure, a wild animal man who was tamed by civilization. Pan was of the same camp, as was the figure of the popular forest dweller Robin Hood.

For a few years, during my earlier explorations of the brook and the surrounding forest and farms of Scratch Flat, I used to meet a strange little man in the woods who dressed in buckskins and attired himself with furs and feathers. He claimed he lived in the woods but would never tell me where, and he used to make his presence known by whistling like a bird. He was around for a few years and then mysteriously disappeared and has never been seen since.

I was telling Kata about this figure years after he disappeared, and she said he may have been a modern-day version of the Green Man. He had all the characteristics; he dressed in leaves, lived in the nearby forest close to the local farms and isolated houses, and as far as I knew was benign—very like the traditional Green Man, in other words. But I later learned from one of the older residents in the community that he actually had a name and lived at home with his mother. He was a bit off, my source said—kindly, and a little slow—and he used to dress up and go out in the woods. He modeled himself on Daniel Boone, I was told.

· · ·

Not long after the autumnal equinox I took a long, slow canoe trip up the stream. It was a warm day, but the signs of autumn were upon us. The goldenrods and asters were in full bloom, and there were certain days when the heat shimmered off the water and clouds of midges and dancing flies hovered in the upper airs. In the quiet coves, the duck weed had spread in sheets over the surface of the waters, and the arrowheads, blue flags, and water lilies were long past bloom and riddled with the holes and trails of insects. The swifts and tree swallows were still coursing above the marshes, and the red-wings were wheeling over the marshes in ever-increasing flocks. Rusty blackbirds and grackles were gathering, and soon the veritable rain of acorns would begin to fall, along with all the other tree nuts. For those who keep watch on these sorts of things, it was clear that the summer was over.

CHAPTER ELEVEN
THE HUNTER'S MOON

Henry Thoreau in his final years was suffering from tuberculosis and was weak and periodically fatigued. But even so, even in his last years in 1858, he was continuing his daily rounds of nature study, ranging out over his territory almost daily. In October, he made notes in his journals of the onset of autumn, and it is interesting, reading through his accounts, how much earlier autumn was arriving in his time. The flowers set seed earlier, the leaves were turning earlier in the month, the nights were colder, and there were killing frosts in early September.

I, too, have been watching the seasons over the years, although in a far more haphazard way, recording the dates of the first frosts, and the spring arrivals of birds and the flowering of spring plants. When I first moved to the area, the first light frosts would come around the eighteenth of September, and the first killing frost would occur around Columbus Day. Now it is possible to have a light frost by October 5, but not always, and the first killing frosts usually occur around Halloween, although not uncommonly they happen in later November, and one year we did not have a real frost until Christmas.

Down on the brook in early October, I noticed that the beavers had cut fresh wood for their winter lodges. Joe Pye weed was still flowering; flocks of mallards and black ducks seemed more plentiful and the reed canary grass and wool grasses were blooming in the marshes. But the frost held off, and there were even hot days when I would sometimes jump into the brook to cool off.

As long as the frosts hold off, life flourishes along the stream. In fact, some of the invertebrate species are more apparent, namely dragonflies and spiders. The eggs of many species of spiders hatch at this time, and on windy days the young spiderlings crawl out to the edges of leaves and twigs, and pay out their gossamer, silken strands and fly off for new territories.

There is one species of very large spider that prefers to weave her webs under the Beaver Brook bridge. Often I take my boat under that bridge and paddle out into the open water of the bay that has formed behind the falls at the north end of the lake. It's a low bridge, but it's easy to get through in early summer if you duck down. But by late summer there are many large, long-legged spiders under the bridge, which I believe are members of a species related to the fishing spider. Their large webs cover almost half the clearing under the bridge and will entangle you no matter how low you try to duck. Even though I know that in reality most species of spiders are reluctant to bite, I don't like being tangled in their sticky webs; in fact, I still have some sort of atavistic repulsion to spiders. But that, I believe, is the result of having grown up in Western culture where the

dark reputation of spiders holds that there are many poisonous arachnids, and that spiders are the associates of witches,
the Devil, and other dark spirits. In other societies, spiders are
benevolent creatures and even helpful.

Mingo, the African slave who lived along the brook for most
of his adult life, would have had a different attitude toward
spiders, for example. He was likely brought over as a child and
held in Curaçao or one of the West Indies slave ports until he
was brought to New England. Mingo was only slightly older
than the two children of his master, and if I have my history
right, the three of them may have worked together in the
orchards, and if Mingo was anything like Tituba in Salem,
he would have shared his African folktales with the children.

In many parts of West Africa, there was a famous trickster
god named Anansi, who was part of the creation of the world.
He was a semi-human spider figure sometimes depicted with
a human face, and he appears in multiple folktales, including
one in which he tricked the creator god Nyame into releasing
stories into the world. Until Anansi fooled him, Nyame had
hidden all the stories in a big box.

As a result of the slave trade, Anansi came to the West
Indies, where he evolved into a symbol of resistance. In many
of the New World folktales, he tricks slave masters and helps
the enslaved people escape. In the following generation of
American-born enslaved children, the tales of a wise trickster
spider character endured, but her name was Americanized and
she became known as Ann Nancy. She was an equally skilled
trickster, along with characters such as the fox and the rabbit.

It's not surprising that spiders should enter in the myths of
disparate societies. The range of spiders is vast; a study years
ago of a new isolated volcanic island in the mid-Pacific found
that the first animals to populate the barren dry land were

spiderlings, who were riding on the high upper winds above the Pacific.

The cultural myths are widespread. Far from Africa, in the American Southwest, another indigenous culture, the Hopi Indians, also had a spider associated with their creation myth. Grandmother Spider, working in tandem with the sun god, Tawa, created human beings and then guided them through an evolutionary series of worlds to become the Hopi people. She herself lives underground in a cave, but she will come forth whenever needed to help her children. In many tales she interacts with the other famous Hopi trickster god, Coyote.

Although they are respected, spiders do not fare as well among the Teton Indians. If a hunter encounters certain species of spider, he would kill it, believing that some evil would befall him if he doesn't. The trouble, however, is that he doesn't want the spider to know that he is the one who did it in, since its soul will go off and tell all the other spiders, who would then seek revenge. So the hunter tells the spider that it was not he but the Thunder-beings who killed him. No one can take revenge on the powerful Thunder beings.

The Cherokee also have a grandmother spider who brought light into the world, which, in earliest times according to the legend, was only darkness. Both possum and the buzzard tried to get light from the sun, but the fur on possum's tail got burned off and the poor buzzard lost all the feathers on his head. But grandmother spider caught the sun in a bowl and then, using her web, spread its light over the earth.

Generally speaking, spiders do not do well in Western culture, although they are not always the evil associates of the Devil. The generic name for spiders is Arachnid, which comes from the Greek myth of the skilled weaver Arachne, who made the big mistake of boasting about her weaving skills. More often than not, pride goeth before a fall in Greek mythology, and Athena took offense and eventually turned Arachne in a spider.

Spiders also prove helpful in some legends and religions in Western lore. Both King David and, later in history, Mohamed were saved by spiders. Each was pursued by enemies, but kindly spiders constructed webs across cave entrances so the pursuing enemies calculated that no one was inside. And it was, of course, no less a figure than Bonny Prince Charlie who hid in a hollow tree trunk when he was under pursuit by Cromwell's Roundheads. While he was in the tree hollow, a spider came along and spun her web across the opening and the Roundhead army passed by. Mere legend, according to historians, but it's a good story.

A spider—or at least a spider bite—lent its name, and possible origin, to a famous southern Italian folk dance called the tarantella. The spinning melody at the end of Mendelsohn's *Italian Symphony*, along with other melodies in classical music, is based on the tarantella, and the dance has been the source of a number of ballet sketches. The folk dance is generally associated with the city of Taranto, although it was found throughout southern Italy as well as Sicily.

The frenzied and tiring spinning dance was believed to be caused by the bite of a tarantula. Whole communities would become simultaneously affected, however, and would dance frenetically in circles, holding hands until they dropped from exhaustion.

Recent scholarship suggests that the wild dancing may have been caused not by a spider bite but by the presence of the hallucinatory ergot fungus that periodically affected rye crops. There is a theory that the bewitched girls of Salem, Massachusetts, may also have ingested infected rye crop.

• • •

The other late summer invertebrate that abounds at this time of year is the dragonfly, which is common in the marshes of Beaver Brook starting in late July. Dragonflies are equally widespread, and are found on every continent save Antarctica, and although they are not quite as famous in lore and legend as the spider, they do have folkloric presence around the world, especially in Asia, specifically in Japan.

Lafcadio Hearn was not only a literary figure and an early ethnologist; he was also something of a naturalist, and in one of his books he provided an illustrated guide to dragonflies, of which there are more than two hundred different species in Japan alone. The nineteenth-century artist Hokusai created a sketchbook devoted solely to dragonflies, and they were the favorites of children who devised various elaborate ways of catching them.

Dragonflies enter the lore of Japan early in its history. According to one legend, 2,000 years ago the emperor Jimmu ascended a mountain overlooking the province of Yamato. He thought the land took the shape of a dragonfly licking its tail, and eventually the whole island was named Ahitsushima, or Land of the Dragonfly. A related legend holds that a horsefly landed on the emperor's arm. Before it could bite him, a dragonfly swept down and ate it. The emperor named the island below in its honor.

Dragonflies are also abundant in North America. Locally, along Beaver Brook, they are common; starting in early summer you can see them patrolling the quiet coves along the oxbows of the stream. For a few years I tried to get a count on the species around Beaver Brook, but finally I had to give up. Identification of dragonfly species involves close inspection of body parts, which, by necessity, involves catching them and—in some cases, in order to keep them still—killing them, something I've never been able to do. Early on in my survey

I quit and satisfied myself with recording the seven families of the orders that occur in the area, including the darners and damselflies, the club tails, the spiketails, and the like. There were a few that I knew from upland areas, including two of my autumn favorites, the large green darner and the red meadowhawk, or sympetron. Green darners are one of the few migratory insects; they make their way south in winter and then return over a period of stages and generations in subsequent springs. By mid-November, the green darner usually disappears, but the red meadowhawk can be found as late as November and even in December on warm afternoons.

Difficult though they may be to identify, our European ancestors were familiar with the common species, and perhaps, needless to say, legends developed about them—mostly bad legends. It was believed that the damselfly or the green darner would sew up the lips of children who lied to their parents, for example, and after the arrival of Christianity, these dragonflies were associated with the Devil. The common names reflect this. They are known as the Devil's Horse in German and the Devil's Needle in the British Isles. The common name "dragonfly" also carries a negative connotation, at least here in the West. Dragons in China and other parts of Asia, by contrast, are creatures of good fortune. In Europe they are evil, dangerous beings. They breathe out a foul pestilential fire and have spikey, dangerous tails and immense curving claws. Furthermore, they often live underground and guard treasures. Heroes of folklore and medieval literature spend a lot of time fighting them, including Beowulf, who was killed by a dragon in his old age.

• • •

The bright month of the Hunter's Moon, which is characterized primarily by the multicolored array of deciduous trees, ends on the darkest of dark nights—the ghost-haunted festival now commonly known as Halloween.

Nowadays, as with so many contemporary holidays, the festival has become commercialized. But when I was growing up, the night streets were filled with devils and ghosts, princesses, queens, kings, pirates, and cavaliers, all of them dressed in costumes designed and sewn by dutiful mothers. Now it's mostly robots and beings taken from electronic games, TV, and popular Walt Disney movies, and all the costumes are purchased in stores or through the internet.

But not that long ago in Europe, Halloween was a decidedly non-commercial festival day; in fact, it was a dangerous time of year. The holiday was known as Samhain in Celtic cultures, the end of summer and the day of the dead, and the tradition was celebrated throughout northern Europe. The celebration falls halfway between the autumn equinox and the winter solstice when, according to tradition, the boundary that divides the spirit world from the mundane world of everyday life thins dramatically. Faeries, demons, devils, and trolls break through the veil and walk the land, wreaking havoc unless they are propitiated.

I was personally reminded of this at a very young age, when, usually around Halloween, my old father would recite, by way of a bedtime prayer, an old Scottish ditty that he must have heard from his family:

> *From ghoulies and ghosties*
> *And long-legged beasties*
> *And things that go bump in the night*
> *Good Lord deliver us!*

This was also, however, a period of harvest in Europe. The swine that had been released into the local forests to feed on mast in spring were rounded up, slaughtered, and salted for the coming winter, and plates of food and drink were left outside at night in order to appease the night beings that roamed the land, such as the Green Man.

Celebrants in the past would dress in costumes and parade the streets, and peasants would perform an early version of trick-or-treat, going door to door, tapping on windows and doors, and begging for sweets and baked goods.

This period between summer and winter was a worldwide time of change, and in pre-Columbian Americas it marked a period when long-dead ancestors could join the land of the living for a brief visit. After the introduction of Christianity to the Americas, the holiday evolved into a celebration of feasting and processions and family visits with the living as well as with the ancestors. Families would gather at grave sites for picnics and lay flowers on the graves of their relatives. Nowadays, the holiday features processions with morbid death masks and skulls and skeletons. These are not considered evil or threatening as they are in North American and European culture; however, they symbolize the return of the ancestors to the land of the living.

In this country, Halloween is not only a time of ghosts and devils and witches, it is also a time when the devilish, flying, sharp-toothed evil creatures known as bats are especially apparent, symbolically. How it came to pass that in the West these beneficial players in the balance of nature, as with so many benign creatures such as dragonflies and spiders, should become associates of the Devil is, I suppose, unknown. But bats certainly are counted among all the other unpopular animals. They fly by night, they have a traditionally ugly, squat face, and if you come near them in their roosts and wake them they will lift their heads and bare their sharp little fangs. Nonetheless, in cultures around the world they are honored. From ancient Babylonia to Madagascar and east to China they are revered, believed to bring good fortune, and even considered gods and demi-gods in certain cultures. Kata tells me that the Queeche Maya believe that the bat Camazotz was a god of the underworld. He lived in a cave labyrinth and assisted the more powerful deities in the grand cycle of ages that was part of their cosmology.

In Celtic lore, by contrast, and also much of the lore of northern Europe, bats were considered messengers of death, and they would emerge at dusk—the witches' hour—and fly with them. And, of course, they are a critical ingredient of Shakespeare's witches' brew, "wool of bat" along with eye of newt and heart of dog.

Be that as it may, in spite of my cultural background, I have always liked bats. When I was a rambling boy of ten or twelve, I used to lure them down out of the sky behind my Aunt Nannie's horse barn in Centreville, Maryland. They would emerge at dusk and flit through the green evening en masse in those distant days, and I would throw stones into the air and watch the bats follow them earthward in the mistaken belief that they—the bats—had located an exceptionally large insect. Bats also used to appear over the marshes of Beaver Brook, and I would see them above my garden, which is located on a rise above the marshes. Now, as a result of a devastating plague of fungus in the local bat communities, the sky is empty.

The bat's evil companions, vampires, actually predate association with bats in European folklore. In fact, the common name "vampire" for the various American bats of the genus *Desmodus* would not have been known in Europe until the sixteenth century, whereas vampire legends predate the arrival of Christianity in Europe. Vampires back then had no relationship with bats: they were supposedly spirits of the dead who had for one reason or another restless souls or were revenants, the undead. They lived in their deceased state by day, but on certain nights they would get up and walk around the villages and the countryside where they were associated with evil deeds, the most common of which was their habit of consuming the blood of the living. Certain demon gods of the ancient world were also revenants and were almost always female. The Hebrew demon witch Lilith was one of these, and she had counterparts all through Europe.

There are also legends of revenants in Asian cultures, although they are not specifically blood drinkers. And there is a thing called the chupacabra that haunts the forests of the Americas. The chupacabra is a blood sucker, but it does not often attack people. "Cabra" implies that it favors the blood of goats, but it also attacks dogs and cattle, apparently.

I met a man years ago at an ecological lodge in Puerto Rico who claimed to have seen a chupacabra. He told me it was a human-like creature, a two-legged being, green in color, he said, and it had long, pointed, upright ears and large eyes. He said it lives in remote forested valleys and is active only at night. He and a group of five friends had heard rumors of one living in a valley near their village, so they went out one night to try to catch it. They formed a circle around the spot where they thought it might be located and slowly closed in, shouting and clapping their hands.

"We saw it briefly," he said. "Green, about four feet high, spindly arms and legs and large ears. But it jumped when we approached and somehow escaped."

• • •

Nowadays, by the end of October the first frosts are likely to settle. This used to be something of a memorable event for me when I was young. There was money in the town in which I grew up, and although we were not rich, we lived just across the street from an expansive property with a brownstone estate, and grounds laid out by the firm of Frederick Law Olmsted.

My bedroom faced a sloping grassy mead that swept up from the street to a wooded hilltop, and in late autumn I would wait for the arrival of Jack Frost, a character I thoroughly believed to exist. And then in the middle of one cold night he would show up. I would look out the window in the morning and the normally greenish brown expanse of ground would be feathered in white, brittle crystals.

By this time, via my parents, who were old enough to still have connections with world mythologies, I knew of Jack Frost, as did most children, but I had also heard stories of the Frost Giants of Norse mythology, and also stories of the Snow Queen and other frosty legends from my mother, a school teacher, who also taught, along with Greek and Roman mythology, Norse tales.

The Norse trickster god Loki lived among the Frost Giants, who were huge icy creatures who lived in an alternate world known as Jotunheim. According to Norse mythologies, at Ragnararokt, the twilight of the gods, the world collapsed in a huge apocalyptic battle and most of the important gods, including the Frost Giants, died (only to be reborn later into a new world).

The arrival of the first frost in autumn is still a major event among gardeners. Frost and the shortening days of winter and the decline of sunlight is indeed the end of a world. To the ancients, without the intercession of the gods, there was perhaps no guarantee that the earth would ever flower again.

This is also the time of year when the local black bears are fattening up in preparation for hibernation, although bears do not exactly hibernate, technically; they just drop into profound drug-like sleep. True hibernators, such as woodchucks and woodland jumping mice, are almost dead for three months of the year. The woodchuck's body temperature drops to about thirty-seven degrees and its heart slows to three or four beats per minute. Bears just sleep. And the female wakes up enough to give birth to her tiny cubs and then goes back to sleep while they nurse.

She and the cubs emerge in spring and begin ranging over her territory. But by autumn, she is ready to mate again and chases off her poor, loyal cubs to make their way.

It is this time of year, around Beaver Brook, when one is most likely to experience bear raids on bird feeders. The young ramblers are out searching for new territory.

Early one autumn morning some years ago, I went down to the kitchen to make coffee and noticed a dark thing just outside the screen porch. This was before coffee and I paid no attention, but I looked out again and saw that the form was a bear. It had reached up and pulled down a bird feeder from the exterior porch rafter and was now leaning casually against the screen porch rails like a degenerate Roman emperor, feasting on the spilled seed.

It came to me that this was perhaps one of the yearlings of a second-generation bear that were ranging out looking for new territory.

I am, of course, fond of bears, and wish them no ill, so I quietly opened the porch door and stepped out to watch it feed. Bears don't have the best eyesight, and she—I believe it was a young female—hardly noticed. The only reaction occurred a few minutes later when our neighbor down the hill let her dog out for the morning. The dog always issues a series of barks when she first exits, and when she did, I saw the otherwise contented bear, who had been casually lapping up seeds, snap her head around, raise her snout, and sniff the air. Then she went back to breakfast.

In time she finished and walked off down a narrow path between the house and a border of rhododendrons. I stepped out onto the terrace to watch, but for some reason, still unaware—or at least indifferent—to my presence, she turned and started back down the path toward me. I'm sorry to report that I used strong language and in so many words told her to get the hell out and go back in the woods and eat grubs like a good old-fashioned wild animal.

Instead of leaving, she charged.

I had learned enough bear lore to know (or at least hope) that this was a false charge. She ran forward a few steps, slapped her forelegs forward like a playful dog, and huffed at me, more like a barking puppy than a wrathful bear. Then she turned and fled down the path and suddenly changed course, crashed through the rhododendrons, scrambled up a hickory tree, and looked back at me, piteously.

I had also recently learned that bears, like dogs, can understand the tone of human language. My rough talk was perceived as a threat, so I changed my tone, and as would any good Native American hunter-gatherer, I spoke kindly to the bear. I apologized and suggested, politely, that it was probably better for her health to feed on wild things rather than store-bought seeds intended for lowly birds. I had actually found a wild honey bee nest recently, and I told her how to get there. She seemed to understand, and she slowly backed down the tree, wandered off across the garden, climbed the back wall, and went off into the woods.

With the return of the native forests in New England, bear populations have increased in the past thirty years. They were rare in the Beaver Brook region, but now some of the common subjects among neighbors and acquaintances are stories of bear encounters, usually as a result of bear raids on bird feeders.

Among traditional hunting cultures in the northern hemisphere, talking to bears is, or was, not uncommon. Bears were considered close relatives of human beings, partly perhaps because they have a plantigrade footprint, as do humans, and

they stand up on two legs from time to time. Traditional hunters among the local native tribes would commonly apologize
to bears after killing them and sometimes even offer them
tobacco to appease their spirits. And there are, as previously
stated, many folkloric tales of human and bear marriages—
always between a human female and a male bear. I don't know
of any tales of men marrying female bears.

Much of this bear lore was collected together in a seminal paper by the anthropologist Irving Hallowell tilted "Bear
Ceremonialism in the Northern Hemisphere." According to
Hallowell's research, much of which has been expanded by
further studies, bear lore reached as far south as Greece and
even into Turkey and the Middle East where the now rare
Syrian bear once roamed.

Not far from Beaver Brook in the 1980s, an archeologist
turned up a bear skull at a dig, which was buried with a flat
stone above its head. This was a tradition among bear-hunting
peoples throughout the northern hemisphere. The Ainu, a
Caucasian-like tribe living on the northern island of Hok-
kaido, in Japan, also had great respect for the bear, and after
much ritual (including a "marriage" to a local woman) would
sacrifice the bear and bury it, laying a flat stone above its head.

Bear lore was common also in Western cultures. It appears
in the story of Goldilocks and the three bears, and, in a more
significant and perhaps more ancient tale, a bear figures as a
sort of surrogate husband, as with the Ainu culture, in the tale
of Rose Red and Snow White. In that story, a bear is taken in
by a human family one winter, and the two daughters groom
him and comfort him throughout the season. In the end, the
bear turns out to be a prince who was cursed by a wicked dwarf.
He marries Rose Red, and his brother marries Snow White.

There is even a body of bear lore in ancient Greece. In Arcadia, the nymph Callisto was one of the followers of Artemis, the

hunter goddess, who had a troupe of loyal virgin huntresses. Zeus took a liking to Callisto and had sex with her. She became pregnant and gave birth to a son, Arcas. When Artemis discovered the union, she was enraged and expelled Callisto, but Hera took pity on her and turned her into a bear. One day her grown son, Arcas, was out hunting and spotted a healthy bear and drew his javelin, unaware of the fact that the bear was his mother. Zeus took pity, swept in, and cast them both into the sky, where they became Ursa Major and Ursa Minor.

Artemis later got her revenge by never allowing them to descend into the sea to drink. They are, in fact, circumpolar, the only constellations that do not dip into the ocean.

In later tales, the two bears became rulers of the Arcadians, who were known as the "bear people."

A bear figure also appears in the myth of Jason and the Argonauts. During their quest for the Golden Fleece, the Argonauts met with the powerful virgin huntress Atalanta, who was raised by a bear. Her father, King Iasus, wanted a son, and when a girl was born instead, he took the baby to a mountain top and left her to die. But a she-bear found her and suckled her and cared for her. When Atalanta grew up, she learned to hunt and care for herself in the forest, as would any normal bear. She was a great fighter; one day, when two centaurs attempted to rape her, she killed them with ease. Eventually, she took the oath of virginity and joined the bear-like hunter girls of Artemis.

There is an ancient cycle of worldwide folktales of a figure known as the Bearson. In this story, which also occurs among Indian tribes of the American West, a bear marries a local woman and the two take up residence in the bear's cave. In time, she gives birth to a son. But the bear husband learns from a soothsayer that the son will grow up and kill him. So the bear rolls a huge rock in front of the cave mouth and imprisons his wife and child. He goes out to hunt every day and brings food back to his little family. Eventually, the bear's son grows into a very strong young man. He rolls the huge rock aside one day and escapes, freeing his mother. He later encounters his father and, true to the prediction, kills him in a fight.

Over the years, the Bearson was incorporated into other folktales. He became an infamous trickster, the hero of many pre-Hellenic Greek folktales. Scholars have argued that the Odyssey is in part a collection of folk tales in which the hero, Odysseus, replays the roles of the traditional trickster of lore, along the lines of the Bearson. The most direct link to this would be the story of the giant Cyclops, Polyphemus.

When he reached the island of Sicily after the fall of Troy, Odysseus and his men unwisely took shelter in a cave belonging to the one-eyed Cyclops, Polyphemus. When the giant came home from tending his sheep that night, he discovered the sailors and rolled a huge boulder in front of the cave mouth and proceeded to eat a few of them. After dinner he asked for the name of their leader. The wily trickster Odysseus announced that his name was Otus.

In ancient Greek, the word "otus" means owl, the symbol of wisdom and Athena. But it also means "nobody." "My name is Nobody," he said, in effect. Those who remember the story will recall that after the crew managed to blind Polyphemus, all the other Cyclops, hearing his bellows, came to the mouth of the cave and asked what was wrong. "I am blinded," Polyphemus called out. "Who blinded you?" they asked. "Nobody," he answered.

His fellow giants departed the scene and Odysseus, aka Nobody, or Wise Owl, and his men escaped by holding onto the underbelly of the sheep when Polyphemus rolled the boulder aside to let them out to graze. In the Native American version of this cycle of myths, the bear is not so bad. He does marry a human female, but their children grow up to be strong leaders. Tom Doublet may have been familiar with this story, or some variation of it. In the pre-contact period there was a local folk tale about a shaman named Paukawna who could turn himself into a bear. He lived on for centuries, either as a man or a bear.

The legend, or at least a version of it, carried into the nineteenth century and was recorded in a somewhat questionable history of the Scratch Flat region. In this version, around 1812, a free African American man named Johnnie Putnam was living in a cabin in a woodlot on a ridge on the eastern edge of the tract. One winter night his dog began whimpering and scratching at the door to be let in, and the next morning Putnam noticed the tracks of a large animal circling the little house. He followed the tracks and saw the huge form of a black bear resting half way up a hemlock tree in the woodlot.

Putnam ran to the house of Joel Proctor, the owner of the woodlot, and the two of them returned to the woods. The bear was still there, apparently sound asleep, whereupon Proctor raised his gun, took aim, and shot the bear, who tumbled dead at Johnnie Putnam's feet. Proctor told Johnnie to wait while he got his ox team hooked up to drag the bear's body home, and Johnnie sat down, leaned against the tree, and took out his pipe for a smoke. While he waited, the bear began to stir and then suddenly leaped to its feet and began stumbling around the hemlock grove, sometimes raising itself on two feet and yowling and screaming in a most hideous manner until finally, spent, it dropped again to the ground and died.

This incident somehow undid Johnnie Putnam. In some versions of the tale, he joined up with Proctor and went off to fight in the War of 1812. He came back a changed man and settled back in his cabin to live out his days in solitude. He never spoke again about the bear incident until shortly before his death, when for whatever reason he retold his tale with a detail he had never shared, not even with Proctor.

"Wasn't no bear died that day," he is recorded as saying, "was a man."

My sources on some of this bear lore come to me via Kata, who in her many-faceted career in Native American studies took a course at the Harvard Peabody Museum that resulted in a prize-winning essay on bear images in the art of the Pacific Northwest and Plains Indians. In that paper, she collected art depicted on teepee covers, totem poles, shields, and masks, as well as a host of Native American bear tales. Several of the images she came across depicted elaborate circling bear dances, sketched by early explorers in the American Prairies. The bear, as in cultures around the world, was always considered part human.

• • •

There was a freakish early snow one October night on Scratch Flat. I went out in the morning that day and noticed that the heavy steel pole hanging one of my bird feeders had been bent double. The feeder itself was crushed, and there was spilled seed all around in the fresh snow surrounded by a set of plantigrade tracks. A savage man had come out of the thick woods beyond the garden wall, destroyed my feeder, and then wandered off back into the woods.

Tom Doublet, were he with us now, I daresay could tell me exactly what manner of man he was.

THE BEAVER MOON

The world does not end abruptly in autumn. After the first frosts there are periodic days of Indian summer when the warm sun returns and the musky smell of old leaves and goatsbeard fills the air. Down on the marshes of Beaver Brook on warm sunny afternoons you can still see the little red meadowhawk dragonflies. I now and then catch glimpses of pickerel frogs or leopard frogs—I never can really identify which is which. I can also hear from time to time the ringing call of a spring peeper, as well as the plaintive whistle of the white throated sparrows. Water striders are still active in quiet backwater pools along with whirligig beetles. And if you look down in the now-cleared waters, you can see turtles swimming as well as schools of dace and sunfish.

winged blackbirds have usually departed by this
huge mixed flocks of blackbirds sometimes sweep
marshes and the treetops. Beavers retreat to their
lodges; muskrats finish their mound building and then, some
grey day later in the month, with lowering clouds and a hint
of moisture in the air, you may see a few flurries of snow.

"Winter is a-coming in, Lude sing goddamn . . . ," as old
Riggs always used to say, quoting Ezra Pound.

• • •

In early November 1858, Henry Thoreau was still making his
rounds of the Concord intervales, recording the setting seeds
and the remnant flowers of the woods and fields and describing, as usual, the range of tints on the last trees. At one point,
he reports hearing the last call of a cricket. He couldn't find
it, but he heard it.

On warm nights in November, you can still hear the much
slowed down, steady pulsing of the snowy tree crickets and the
last sad scraping of one or two katydids. And indoors, by the
hearth (for those happy few who still have hearths) you can
hear the bright chirping of a field cricket. But after the first
frosts, gone are loud full choruses of the night-singing insects.
One by one they slow and then sign off until the night, once
so full of the stridulating noise of life, is silenced.

The whole affair of the slow, step-by-step departure of the
singing insects reminds me a little of Haydn's *Farewell Symphony*, when in the last movement the musicians rise, one
by one, and leave the stage. The second violins are the first,
followed by some members of the first violin section, then a
few woodwinds, then the violas, more woodwinds and brass,
the cellos, followed by the bass, until finally all that is left are
two first violinists.

And then they too depart, leaving the stage dark and empty and silent.

• • •

As did most of the people around Concord in the 1850s, Henry Thoreau did not always favor the local Irish. There was, for example, an old shanty man who lived near Henry's cabin at Walden who Henry described as shiftless, and he also used to make fun of Emerson's Irish gardener. But over time his attitudes began to change, and unlike most of his local neighbors, he came to appreciate the Irish workers who lived in Concord. In fact, on one occasion, when a local family man attempted to cheat his Irish worker out of funds he had won at a town fair, Henry took up a collection for the Irishman.

The Irish families lived in deplorable conditions, mere shanties in most cases, and in spite of their willingness to work and their general good humor—as Henry's friend Margaret Fuller pointed out—locals were unwilling to help support them. These upright Concordians, by the way, were the same people who were strong supporters of abolition.

During this same period, the late 1840s, there was an Irish family who lived for a while in a shanty on the eastern side of the Beaver Brook. There are scant records of the family, but whoever they were they would have brought along with them all the legends stories and beliefs from the old country, one of which involved an animal spirit called the Pooka who always stalked the land in the month of November.

Long ago, before Saint Patrick came to Ireland, the Pooka was well established in Irish folklore. He lived among ruins in lonely mountain retreats and he was a creature of nightmares who would emerge from his solitary lair in the remote hills every year in early November. The Pooka was a terrible beast and, as were so many animal demons of this sort, he was also a shape-shifter. He would appear in many forms. Sometimes he was a wild horse who could speak in a human voice. Sometimes he was a goat, sometimes a donkey, or an eagle, or a bull. But in whatever form, he was able to predict the fortunes of those who dared to confront him.

There are no remote mountain crags anywhere near Beaver Brook, but there is a high hill just northeast of the marshes, and there, in the forested slopes and the orchards below, there were living a number of historical and semi-mythical beings. One of the most intriguing stories of the land near the hill is the existence of an image of a fourteenth-century sword-bearing armored knight, chiseled into a flat slab of granite just north of the hill as a memorial to the fallen knight. The image is believed to be a man associated with a seafaring Scotsman named Hugh Sinclair who, two years before the arrival of Columbus, was blown southwestward in a storm and ended up ascending the Merrimack River and Stony Brook, which is the northern branch of the Beaver Brook. On the way up the hill to have a look at the surrounding countryside, one of Sinclair's knights died. The crew cut a memorial tribute to him in the stone—or so the story goes.

The knight—if he indeed existed—was perhaps terrified by a vision of the nightmarish Pooka in one of his various beast forms, or he may have had a vision of another wild creature believed to haunt these hills, a ghostly white stag. This white-tailed deer, a huge stag with a fine set of branching antlers, seems to have had a presence in the region even into the twentieth century. He was seen by local hunters over a period of years in the 1970s, but even though they believed that they had aimed true on those few occasions when they were able to get a shot off, the stag bounded off, unharmed. His last appearance, back in the 1990s, took place in the yard of an old woman who lived on the north side of the hill beyond Beaver Brook.

In November, while she and a circle of friends were sewing in the living room, the stag approached the house and peered in the window. When the women spotted him, he spirited off through the orchards.

The woman's gardener had also seen this stag on several occasions. He said that later, in the December hunting season the year before, a hunter encountered him and managed to get a shot off but somehow missed.

Legends of mystical stags are legion; in fact, it is said that the stag is the symbol of more Christian saints than any other animal. One of the earliest of these involved a pagan Roman soldier named Placidus, who, out hunting one day, came upon a magnificent white stag. He drew back his spear to make a cast, whereupon a shining cross appeared between the deer's antlers and the stag spoke: "I am the Redeemer," he said. "Whyfor dost thou persecute me?"

Placidus fell to his knees and converted on the spot and changed his name to Eustace. Conversion does not seem to have helped Saint Eustace, however. He subsequently lost his property and was later executed along with his family by the pagan Romans.

It is said that a white stag also assisted the Christian warrior Charlemagne by guiding him through the Alps. But long before the advent of Christianity, the stag was an important figure in world myth. He appears in Indian tradition as a horned god associated with Shiva. He was a fellow traveler of Artemis in Greek mythology, and he was an important figure in Celtic legend and religion. In this form, he is depicted as a man with a head of antlers, but he is a companion of all stags and is believed to be a god of nature. In fact, the deer is often believed to be something of a messenger between wilderness and civilization in many cultures, and not always a benign one.

One of the few human-like images painted on the cave walls of Les Trois Frères in Ariège, France, is an image of a stag man known as "The Sorcerer" or the "Horned God." He has the upper body of a man but deer-like legs, and a full set of antlers and a couple of other animalistic attributes, and he is crouched over with his head facing forward. The drawing is thought to be about 14,000 years old. Scholars are not certain who or what he is, but one theory holds that he is perhaps some overarching god-beast who may be an intermediary between human hunters and the animals.

Kata told me that the Kiowa people lived in terror of a vaguely similar character, a numinous deer woman who would sometimes appear at ceremonial dances and, in the form of a beautiful woman, lure young men away from the dancers, and then, once off in the wilds, transform herself into a ruthless doe and trample her victims to death. Kata said she once met an old man who heard it from another man who heard it from a man who was at a powwow in the late 1950s when the deer woman appeared. One young man mysteriously disappeared from the frenzied midst of the circling dancers that night and was never seen again. Fellow dancers thought they saw, briefly, the figure of a woman joining the dance and reported that she had hooves instead of human feet.

In these parts, the Eastern Woodland people also had a deer woman, but she was more benign and was associated with fertility and was a welcome figure. That did not keep the local native hunters from killing deer. Venison was an important staple of the Eastern Woodland tribes. In autumn, the meat was dried in strips to make pemmican, one of the staple winter foods of the northern tribes. Deer were so important to the native economy that the Indians would sometimes burn over sections of the local forest to encourage the growth of blueberries, which would then attract both deer and bears to the ripe berries.

For all his wanderings in forest and fields, Henry Thoreau only saw a white-tailed deer once or twice. The land in his time was eighty percent cleared of trees, and what stands of trees were left were used as woodlots for firewood and construction timber. The great wild forest of the Northeast that grew up after the retreat of the glacier was thoroughly stripped, but with the decline of farming in New England it came back and now the statistics are reversed. The Northeast is currently eighty percent forest. Except for the Pacific Northwest, it is the most heavily forested region of the United States.

I was walking home from the brook at dusk one damp, mid-November evening through the half-lit forest when for no particular reason that I can remember something stopped me dead in my tracks. There were wraith-like mists curling up from low spots on the wet forest floor, and the air was still and somehow portentous, as if there were some ghostly presence nearby. This is not an uncommon sensation for me, especially in certain sections of the woodlands in these parts, which, according to local legends, is supposed to be haunted. But this feeling was a little different, and after a minute or two I realized what it was. It was the forest itself, the inhuman, dangerous wilderness of the Western mind. The sensation, or belief in the danger of wild places, has been with us since the dawn of civilization.

Just before he began on his epic journey through the nine circles of the Inferno, Dante was waylaid in a dark forest where the straight way was lost. Dwelling in this same forest—somewhat incongruously, given the fact that the event took place in thirteenth-century Tuscany—were any number of dangerous, exotic beasts, including a leopard and a lion, and also a wolf.

The fact that a deep and perilous forest would characterize the outskirts of Hell is not illogical, at least not to the
thirteenth-century European mind. It was here, after all,
beneath a great overarching canopy of ancient trees that
modern humans first forged the myths, legends, and religions
that until fairly recently played such an important part in the
human experience. And from the beginning, folktales and legends make it apparent that the relationship between the forest
and the peoples who lived in or near it was complicated.

The Sumerian epic of Gilgamesh, the first written narrative,
which was set down from the stories of a much earlier era, is a
case in point. In the epic tale, the hero Gilgamesh befriends a
half-man, half-animal being named Enkidu who was a friend
to animals and fed with them in a semblance of a paradise.
Enkidu was tamed, interestingly enough, by a courtesan from
Uruk, who was sent out to have sex with him. After a week of
furious lovemaking he was somehow transformed and gained
knowledge of the world of human beings. But his former
friends, the animals, were now afraid of him.

Enkidu and Gilgamesh, the king of Uruk, became fast
friends, and working together they set out to do battle with the
guardian of the Great Cedar Forest, a terrible monster named
Humbaba, who was sent by the god Anu to guard the forest
against people. The two friends managed to kill Humbaba,
and then, with the guardian defeated, they proceeded to cut
down the entire forest.

And so beginneth Western civilization.

By the fifth century BC, Plato was lamenting the loss of the
pine forests that formerly covered the mountains of Greece,
and there is good evidence that by the time of the Roman
Republic most of the native forests of the Italian peninsula,
save for a few remote mountain valleys, had been cleared.

But the old legends, and the ancient forest gods and demi-gods from this earlier forested environment, were alive and well in the clear light of fifth-century Greece.

In one myth from early Greek history, a young forest nymph who lived in the wilds of Arcadia was courted by the god Mercury. In due time she gave birth to the god Pan, who, not unlike Enkidu, was half-animal, half-human. He had the face and upper body of a man but the shaggy legs, hooves, and horns of a goat. According to the legends, Pan's mother was so afraid when she first saw him that she ran away. But Pan grew up to become one of the most entertaining and popular of the Greek and Roman gods and served as a sort of intermediary between wild nature and human settlement. The Lord of the Wood, as he was called, haunted the forested mountain passes above the pastured lands of Arcadia and struck fear into travelers who passed through the valleys and ridges at night. You didn't even have to see him to know he was there; you could feel his presence in the form of "panic."

Pan is the only Greek and Roman god who actually died; all the others merely faded away. At the time of the birth of Christ, or, in some versions of the legend, at the crucifixion, pilgrims traveling in Italy heard a thundering voice echoing through the mountains and valleys, crying out that the Great God Pan was dead. This news was followed by wails of lamentation as all the nymphs and dryads and the fauns and satyrs, and all the old demigods of trees and springs and groves, retreated to the mountain hideaways to bide their time. After that, it is said that the springs fell dry and the famous oracles no longer prophesied accurately; their keeper, the Lord of the Wood, was dead.

Or was he?

One would think with all my woodland wanderings and my interest in bear spirits and the like that somewhere along the way I would meet some demon or wood spirit in my travels, but in fact, I've encountered such a being only once and that was when I was abroad in an ancient land.

I was staying at a goat farm in the tiny town of Cros in the Cévennes mountains in France, and every day I would pack a tranche of bread, some onions, and a can of sardines and set out to explore. On one of those days, I came upon a ferny ravine with a bubbling stream running through it. It was hot, I had been walking all morning, and so I settled here, cooled my feet in the chilly waters, and ate my lunch.

Somewhat fatigued from the morning's hike, I found a mossy patch and laid down to rest and, in time, fell half asleep. In my drowsy state, I could hear the languorous babble of the waters, the occasional bursts of wind, and the distant clang of goat bells. Maybe I fell into a deeper sleep, and maybe I dreamed what followed, but I heard the distinct rattle and clatter of loose stones sliding down the rocky scree on the other side of the stream. I opened my eyes and I saw, peering through the brush on the other side of the brook, a hideous bearded face, a shaggy head crowned with the curling horns of a goat and what appeared to be the body of a man.

I sat up abruptly, and as I did so, I heard more clearly now the rattle of cascading stones as the goat, or the goat being, fled up the slopes.

This all happened in a matter of seconds, and I sat back stunned and dumbfounded. I was wide awake by that time and, rational being that I am (or am supposed to be) I decided I either dreamed the vision just before I woke or it was a stray billy goat who had come to the brook for a drink. But the human-like form persisted, in the way in which sharp, clear dreams hang on well into waking hours. This was, after all, the Cévennes, a region rich in Roman ruins, and everywhere I went on my ramblings I would come across evidence of past cultures, reaching all the way back to the Paleolithic era. If Pan were to haunt anywhere on earth, it would be here, in the isolated mountains, ravines, and valleys.

Fortunately for the older pagan deities, word of Pan's demise never reached as far north as Germany, France, and Scandinavia, where the forest-dwelling tribes still held sway. Many of these cultural groups in this region traced their origins to a version of the great World Tree, the Yggdrasil of Norse mythology, an immense sacred ash tree from which all life sprang. Most of the pagan gods, both the good ones and the bad ones, were forest beings in these regions; and the priests and ovates of their religions were inclined to make sacrifices to trees to keep the trees and their associated gods happy—or at least to hold them at bay. The Druid cults in particular were known to worship in oak groves and at sacred forest springs.

In Greek and Roman mythologies, and also in the forest cultures of the Germanic tribes, the host of wild godlike beings inhabiting the woods and mountains was abundant. The one-eyed Cyclops had their origins in the forest, along with the centaurs, the wild, horse-bodied humans, also goat-footed satyrs, fauns, and silvani, who protected sheep and goat flocks as well as nymphs and the tree-dwelling dryads, and sileni, the powerful chiefs of satyrs.

Pan and the sileni were believed to be associates of the Greek god Dionysus, the god of wine and fertility, and would make their presence known in the ecstatic and debauched Dionysian festivals, along with the erotic but murderous women known as the maenads, the followers of Dionysus.

By the fourth century AD, as Christian proselytizers moved northward converting the conquered tribes, these once-powerful deities began to diminish in the eyes of their former worshipers, although it took a while for them to decline altogether. One of the prime targets of the Christian missionaries

was the environment where these gods once dwelt. And in
725 AD the aforementioned Saint Boniface, as if in remem-
brance of Gilgamesh, celebrated the Christian victory over
the German woodland tribes by personally cutting down the
huge sacred Donar Oak, thus destroying the heart of the old
heathen beliefs.

But the old forest spirits did not die easily. In particular Pan,
or someone very like him, continued to appear in the night
forests beyond the villages. Catholic bishops had the wisdom
to transform Pan rather than attempt to deny him, however.
It was a known fact; they pointed out to the ignorant heathens
that the enemy of God, the Devil himself, had cloven hooves,
a set of goat horns, and a pointy little goat-like beard. Even
that strategy didn't work entirely, though. The ancient beliefs
were perhaps too deeply engrained in the European psyche to
disappear in so short a time.

Even into the sixteenth century in rural areas, peasants
would leave milk and grains and fruits at the edges of the
cultivated fields as offerings to the Green Man. As long as he
was fed, the Green Man would help little children and guide
lost night travelers out from the dark forests. Ironically, given
his pagan roots, the Green Man's image, a full-bearded, Pan-
like character who wore a headdress of ivy and fruits, became
a popular icon. His likeness can still be seen carved into pews
and pillars of medieval and Renaissance churches.

A further irony is that the long aisles of the darkened interiors of the great twelfth-century cathedrals of Europe should so
closely resemble an old-growth forest. Rank on rank of straight-
trunked pillars soar heavenward in a subdued, raking light into
the mysterious canopies of beams, cruxes, and trusses. More
to the point is that in more recent eras, the terminology used
to describe the beauty of ancient woods reverses the metaphor
and makes use of the term "cathedral," to describe old growth

forests—sometimes officially, as in the Cathedral of the Pines in Rindge, New Hampshire. Even closer to the idea of a cathedral forest is the architect E. Fay Jones's unique Thorncrown Chapel in Eureka Springs, Arkansas, a soaring wooden structure with 425 windows that is thoroughly integrated with the surrounding woodlands.

The American poet William Cullen Bryant more or less summed all this up:

"The groves were God's first temples," he wrote.

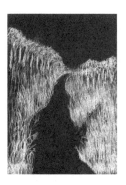

CHAPTER THIRTEEN
THE COLD MOON

There is a certain slant of clear light in December that settles over the snowless landscape early in the month. The hills and fields have gone to brown; the marshes of the Beaver Brook take on a tawny lion-coat color, and the waters of the stream, now clear of aquatic vegetation, run black against the subdued colors of the cattail walls. Nights are chill and frosted, the Cold Moon is sharper in the clear night; Orion rides in the southeastern sky and, for whatever reason, the local coyote packs set up a chorus of ghastly yips and yowls and howls at two in the morning when the full moon casts tree shadows across the gardens.

Down on the brook, as if to match the landscape, a late-migrating hooded merganser floats by, its white hood and black body contrasting the monochrome background. The herds of white-tailed deer collect in the deer yards and wooded clearings, grey-coated now in winter pelage. No summer sounds, no insects or bird calls, save for a lonely semaphore of a gray and black chickadee and twittering flocks of juncos.

The bats have migrated or have secured themselves in sheltered bat caves or in the barns or the attics of old houses; the resident land mammals have gone to ground in dens and shelters and tree hollows only to reappear in fine weather, if such a thing were to come to pass. Deep under the earth, the chipmunks have constructed complex labyrinths of rooms and dens, some for storage, some for sleep, and the woodchuck is all but dead to the world. Bears, too, have quit the scene, asleep in rock shelters or under tree root caves or secluded brush dens. And down on the brook the muskrats have built bankside lodges of mud and reeds and grass stems, and the beavers have constructed elaborate rooms and chambers in their lodges, some for feeding, some for sleeping, and they have anchored branches and saplings in the bottom of ponds and streams and will feed on them throughout the winter.

And all the while, the Great God Sun, the creator of all things on Earth for so many cultures, is declining. These are barren, empty days when one could believe that all life has ceased. Time is dead, a stillness pervades; it is a sleeping kingdom waiting for an uncertain springtime god to burst through the proverbial thorn hedge.

And then, finally, on the twenty-first, the winter solstice arrives, the longest night of the year and the return of light.

• • •

Henry Thoreau always seemed to have trouble with his health in winter, and after 1858 he began to get worse, enduring periods of weakness and coughing spells. In the fall of 1861, he caught a cold that turned into bronchitis, and although he would recover enough to get out from time to time, his cough worsened. Things slowed down. His journal entries shortened and sometimes disappeared for days on end. And then after years of recording his thoughts and excursions and natural history notes, he made his last entry on November 3 with a comment on the value of sharp observation. He had completed what critics consider his finest but least read work.

By December he was housebound and continued to decline over the course of the winter, although he attempted to carry on with his readings and visits from friends. Slowly, day by day, he grew weaker. Friends and family continued to visit, recognizing that his end was near. He was suffering, but he did not apparently fear his oncoming death. His friends and family members worried about his relationship with God. One asked how he felt now about Christ. He told her he cared more about snowstorms than he did about Christ. His old Aunt Louisa asked him at one point if he had made peace with his maker. He answered that he did not believe they had ever quarreled.

Henry never was a churchgoer. One could argue that if anything he was an animist; as did the local Concord transcendentalists, he saw a spirit—if not a singular God—in all things. He had his own version of the Nicean Creed:

"I believe in the forest, and in the meadow and in the night in which the corn grows," he wrote.

He languished until spring and died, ironically, in May, just as the warblers were returning and his beloved natural world with all its hope and rebirth was greening up.

His last spoken words were a fitting epitaph for a man so deeply immersed in wild nature. "Moose," he whispered. "Indian."

· · ·

Here in Thoreau Country, at some point in early winter, a light snow will drop, and if you go out in the early morning, you will see that the trees have been stitched together with mouse tracks and the lines of fox and coyote trails, the patterned tracks of rabbits, and the paired tracks of squirrels, and even the rounded track of a bobcat. Also, the tracks of what appears to be a large man, walking barefoot in the snow—a bear.

I also notice otter tracks and their fish-scale-laden droppings on one side of the bank. It seems that periodically over the course of the year they, too, enjoy resting at what I consider to be "my" spot. Earlier in December I had also seen an otter slide on the bank, a muddy chute leading down into a tangle of button bush and cattails.

Among the Eastern Woodland Indians, the otter was considered the winter bringer. Why this should be is unknown since otters are active all year. It may be that their tracks are more evident in winter and the fact that otters often keep holes in the ice above warmer springs all through the winter.

Legends of otters are widespread and seasonal associations are common. Among the Northeastern tribes there is an otter-like creature known as the water-panther, a powerful beast that has control over all the other aquatic animals. It is a dangerous, dragon-like animal, and no doubt Tom Doublet and his family feared the beast even after they were Christianized. As Kata used to say about the Christianization of the local tribes, they accepted the new religion only because it was easier. They didn't have to go around propitiating deer and bears and thanking the Corn Mother. They could just pray to one god of everything. Bad spirits like the water-panther were also easy to fend off— they were the work of God's arch enemy, the Devil. Pray and you will be safe.

Kata Grant, my source for much of this lore. She was in some ways an eternal student of things—basketry, wildlife, plant lore,

and especially Native American cultures and their relationship
to nature. Unlike so many white Americans, however, who have
accepted and even practiced Native American traditions and
religions, she was well enough read in the field to question the
standard belief that the native people of the Americas lived in
peace and harmony with nature. In fact, she says, they lived in
fear of a host of woodland demons and monsters, and Indian
hunters and trappers, without apparent qualms, nearly wiped
out the beaver and the white-tailed deer at the behest of the
European traders. She also likes to point out that one theory of
the great die-off of the mammals of the Pleistocene era is based
on the fact that the ancestors of the Indians, the Paleo Indians,
were responsible for the extinction of the woolly mammoth and
other large game animals. They used to drive the herds by setting
fires and would run them over cliffs, then set up camp, butcher
them, and live on the meat until it was used up.

"These people were, after all, homo sapiens, no different
than us," she said, "not an entirely peaceful species of great ape,
in other words. And in the long run, perhaps not that wise."

Kata used to tell me—half in jest, I hope—that the most
intelligent mammal on earth is the possum. "It post-dates the
dinosaurs by only a few thousand years," she said, "and it has
shambled through another hundred million years or so, mind-
ing its own business and doing its possum thing, giving up
when threatened and rolling over dead, only to rise again after
danger has passed."

She might have said the same about turtles, except that,
unlike the possum, they are currently threatened in the Northeast. I actually happened to see a turtle one December day by
the brook. I went out on a bitter-chill morning after a serious
cold spell and found that the ice had formed on Beaver Brook.
It was clear black ice and it was thick enough to venture out
onto, so I slipped down the bank, tested the ice, and then lay

down in midstream. It was clear enough to see down into the waters, almost like staring through a bright glass window, and I could see, flowing in the current, the long waves of aquatic grassy vegetation. It looked like the hair of some beautiful Naiad or Undine, stretching green against the dark background of the bottom of the stream. One could understand, perhaps, how the myths of water nymphs would arise.

I had often watched the underwater life in this way over the years when the conditions were right. One year, lying face down on the ice, watching the things that pass in the current, I saw a shadowed form swim by. It was a large wood turtle, a now uncommon species in these parts.

The wood turtle may be the species of turtle that carried the primordial world on its back, at least in the belief system of the Eastern Woodland people. Local Algonquian legends hold that the wood turtle can live for 3,000 years, so it is a likely candidate for a creation story.

Early in December that same year I was following the trail from the bank along the western shore to the footbridge when I came across a clump of whitish, vase-shape mushrooms growing out of an old stump. I knew this species; they were winter oyster mushrooms, a not uncommon edible species, so I collected them to carry home for dinner.

Earlier in the autumn, I had found many different fungi along this same trail. It had been an oddly wet spring and there was rain again that fall. In fact, for some bizarre reason it seemed that there was a hard rain every Friday between mid-September and mid-October. As a result, it was a good year for mushroom picking.

One of the common species I found was the classic toadstool, or fly agaric, the *Amanita muscaria*. These are the red-caped mushrooms with white spots that you see illustrated in many children's books of fairytales and animals. But there is a deeper history associated with this species. This particular mushroom is hallucinogenic, although dangerous if you eat too much of it. Among the Laplander or Sami people, and Siberian tribes such as the Koryak, it was used for ceremonial purposes. Shamans would ingest it and, among other sensations, would assume the ability to fly, and also gain knowledge of important lessons on the nature of the universe. There is even a theory that the mushroom is associated with the legends of Santa Claus.

As did many Laplanders, the shamans would ride in a sleigh pulled by reindeer. They would also often dress in red and wear red hats. In earlier days, the Sami people lived in yurts, which would often be sealed in by heavy snows so that the only access was the centralized smoke hole of the yurt, the chimney. During the winter solstice ceremonies, the shaman would dress in red, consume the Amanita, enter into a hallucinatory state, and fly off into the spirit world. Having gained access to the mysteries, he would return and enter into the yurt through the smoke hole, bearing the gifts of his insights.

Reindeer themselves have a strong predilection for the fly agaric, and if left unchecked will consume patches of the mushrooms before the shamans can collect them. Presumably, although of course no one knows, they too may gain sensations of flying. Shamans themselves may have traveled in these flying reindeer sleighs à la Santa Claus.

Mushrooms, mainly the psychedelic species, are a part of rituals and ceremonies wherever they grow. There is even evidence that the Neanderthals may have consumed mushrooms, possibly in association with ritual burial, according to a few archeologists. Skeletal remains in a 60,000-year-old Neanderthal burial site in Shanidar Cave in northern Iraq show pollen evidence of flowering plants and also spore deposits of mushrooms. The use of what is known as entheogenic plants—plants that, when consumed, induce altered states of consciousness—have been found in Neolithic burial sites and documented in ancient Greece and Rome. Flying, time travel, trips into the spirit world and the Underworld, which so often appear in the myths and even in the philosophical theories of the ancient Greeks, may have resulted from the ingestion of entheogenic plants, including psychedelic mushrooms. The Celts are believed to have used *Amanita muscaria* in rituals, and in societies throughout Mexico and in the Indies the sacred drink soma, a favorite beverage of Indra and other deities, as well as priests and shamans, may have contained substances derived from psychedelic mushrooms. The ingestion of mushrooms containing psilocybin is also documented by the Spanish conquistadores, among the Aztecs and other Mesoamerican societies.

Consumption of psilocybin mushrooms spread into the counterculture in the United States in the 1960s and garnered a bad reputation with federal authorities and was declared illegal. But its use in Mexico in rituals is still common, and even U.S. authorities are rethinking its use among psychologists for medicinal purposes.

An American friend of mine, who lived among rural people in Mexico, knew a local woman who may have been a curandera and who used to eat psychedelic mushrooms for religious purposes. She told him that on these occasions she would go off and visit with Jesus. She told him once that she actually danced with Jesus. (My friend told me he had to force himself not to ask her if Jesus was a good dancer.) He himself used to eat the mushroom just to get high, but he never did see Jesus. All he did was laugh at things that were not necessarily funny.

Jesus enters into mushroom myths in other ways. Back in the early 1970s, the English archeologist John Allegro wrote a book that argued that the historical Jesus never existed. He was, the theory goes, the invention of a cloistered fertility cult of pre-Christian monks, who envisioned and invented a perfect spiritual being they called Jesus. According to Allegro, they had consumed *Amanita muscaria* mushrooms to induce their visions. The book was soundly dismissed by Christian scholars and subsequently taken out of print. But more recently his theories have been reconsidered in relation to the use of psychedelic mushrooms in various religious practices.

I once met an old Mayan woman in the ruins of the seacoast temple Tulum, in Quintana Roo, who I suspect had knowledge of local psychedelic mushrooms and may have been a curandera. I was lingering on the temple grounds after dusk, after all the tourists had left, and fell into conversation with the old woman who, even in dotage, was working on the grounds, sweeping up. I tried to talk to her, but she spoke Yucatec Maya and her Spanish was limited. I wanted to ask her if she had ever seen one of the jaguars that were depicted in one of the friezes in the temple wall in real life. We spoke in sign and in her limited Spanish, and what I learned, or think I understood her to say, was that the stone jaguars on the walls of the temple would come alive at night and leave the temple grounds to hunt the forests.

I caught the Yucatec word for jaguar, *ba'lam*, which I recognized from research I had been doing on jaguars in the local religions of Mesoamerica, and she also understood the Spanish word for jaguar, *tigre*.

It's quite logical that the jaguars come down at night, the old woman told me, since jaguars do hunt at night. Also, the Mesoamerican jaguar god is a night spirit in traditional Mayan religions. He is the "night sun" and a common figure in pottery and sculpture and even appears on temple complexes throughout Mesoamerica. He also appears as a spirit guide among the native Indians in the Amazon River basin.

Shamans in the region will consume ayehuasca, a psychedelic drink made from the sap of a local vine, and subsequently transform themselves into jaguars, and in this form gain knowledge of healing methods and are even able to prophesy future events.

• • •

I saw old Riggs early in the month of December one year. I was standing on the footbridge staring out over the brook when he appeared on his way for a walk along the eastern bank, where there is another footpath that follows the brook. I saw him slowly approach around a bend in the trail. He did not look well, a crabbed, wintery old man with a hickory walking stick, his *de rigeur* slouch hat pulled down over his brow. He was watching his step as he plodded along, his eyes on the ground, and he did not see me until he reached the bridge.

We chatted there for a while and then he carried on up a short incline, over the brow of the hill, and then slowly disappeared below the ridge. It was, as it happened, the last time I saw him.

Later in the month, just before the solstice, in fact, I heard from Rosemarie that he had died. It was *enfin*, a good death. He had spent the morning out on the local trails somewhere, and when he came back, he had lunch, announced that he was tired, and went off for a nap in his study. He died there, apparently of a heart attack. I believe he was ninety-three years old.

After that, I took a memorial walk along the brook trail thinking about him and about one of our last conversations earlier that fall. He had joined me at the landing and sat with me on the ground for a while staring out over the marshes. I asked him at one point about my idea that myth and folklore serve, among others things, to connect people to nature and their local environment.

"It's deeper than that . . . ," he said and launched into one of his grand unifying theories.

He claimed that myth, legends, folklore, and, finally, religion itself all spring from landscape. Northern cultures, he said, have seals, caribou, lynx, and wolverines in their pantheon and their folktales, as well as ice spirits and frost giants. Same with the cultures of middle Europe and the British Isles. These forest cultures dwell in complex and biodiverse ecotones and damp greenery and have complicated stories of forest beings, wolves, bears, tree spirits, and water beings. The Greeks and Romans invented a multitude of gods and goddesses of springs and streams, mountains, seasons, and the sky, and earlier in their prehistory when the land was still forested, they had a chthonic, earth-based religion, with the snake as a primordial being. Also a host of sea gods. Then the hills and mountains are stripped of trees, the land opens with wide vistas, and along comes a body of sky gods, Zeus and Hera, Apollo and company.

And then Riggs expounded his bravest theory of all—the birth of monotheism. His claim was that three of the five great religions of the ancient world came into being in the fifth century BCE and that they emerged from desert people—first from the pre-Mosaic Egyptian pharaoh Akhenaten, who selected the sun as the one true god, followed by Judaism, then Christianity, and then finally Islam.

"All desert cultures," Riggs said. "And deserts are extreme. Cold at night and blazing hot by day, limited in colors, singular, black and white, either/or, yes or no, dualistic—an evil Devil and a benevolent God, no deep forests, no running streams or rivers, and so it's quite logical that you would get a, a what?—a limited palette of deities, ending up eventually with only One God."

"Amen," I said.

"Amen indeed," he said, chuckling to himself, as if he didn't actually believe his theory.

Riggs was raised in the Anglican tradition. He said he never was religious, but after his time in the war and his experiences with religion in the ruins of Central America and the murderous Aztecs and Mayan cultures, he quit religion altogether. He died an atheist, and instead of a funeral, later that winter Rosemarie organized a big cocktail party for his old friends in Cambridge.

I remember Riggs told me once that when he was a boy scout in London, they would hold meetings in a Catholic Church. He would often see the sexton of the church there, kneeling at prayer on the cold stone flags in front of the altar. He learned later that the supplicant was an American named T. S. Eliot.

"Who knows what lurks in the mind of man," Riggs said cynically, quoting someone, I think.

It was perhaps a fitting memory for someone like Riggs, who had such an interest in the world of nature and the world of literature and ancient cultures. It was, after all, Eliot's seminal 1922 poem *The Wasteland* that laid bare the loss of myth in Western culture. How is it possible for modern society to establish creative inspiration in the stony desert of a world that has lost touch with the deep mythical underpinnings of its culture? Myth and ritual obviate disorder.

But it isn't quite over yet.

In spite of the crass commercialism of ancient ceremonies and the downright materialism and greed of traditional festival days such as Christmas, there seems to be a slow and quiet rebirth of interest in the original seasonal festival days of deep history. Here in the United States and especially in Europe, groups of celebrants, mostly young people, are attempting to revive the old pagan ceremonies such as Walpurgis Night. There is, for example, a huge Beltane Fire Festival in—of all places—old, grey Edinburgh, that attracts thousands of people.

Even more bizarre is the popular Burning Man Festival that started in San Francisco as an arts festival. The event has since grown so large the organizers had to move the festival to the Nevada desert. No one, except perhaps the original organizers, seems to pay much attention or even know the fact that the central event of the celebration, the burning of the huge wicker effigy of a human figure, is based on the ancient and very nasty Celtic ceremony in which foxes, cats, snakes, and human sacrificial criminals were burned alive in a massive bonfire in a huge wicker, man-shaped basket. Even the brutish Roman invaders were horrified by the celebration.

More benign is the restoration of Native American pow-wows and cultural events celebrated by American Indians as well as nouveau Indian white followers. Fortunately, the celebrants do not seem to have revived the old practice of the former Lakota Sun Rite in which braves would pierce their pectoral muscles with sharpened bone pins and attach them to a central post with rawhide straps and then lean back and even sometimes have themselves lifted off their feet and suspended. Nor—fortunately—has the dreaded Deer Woman who snatches off young men appeared at any of the contemporary Nouveau Indian dance ceremonies.

Coupled with this revival of earlier cultural traditions, there has been a renewal of interest in faeries, a phenomenon that seems to occur in reaction to new advances in technologies.

As a result of scientific discoveries and increased industrialization in the Victorian Era, artists and craftsmen returned to earlier ages for inspiration, including a popularization of faerie lore. There was a revival of interest in Shakespeare's plays *The Tempest* and *A Midsummer Night's Dream,* and a school of specialized faerie painters developed, including the artists Joseph Noel Paton and Richard Dadd. The paintings depicted elaborate, detailed scenes of faerie life in woodland settings.

The movement even influenced the Pre-Raphaelite painters, who drew their inspiration from the Middle Ages.

In our time, starting in the 1980s with the rise of computers and the internet and cell phones, a much-sanitized version of fairy lore (with an updated spelling) developed. These fairies were mostly female and decidedly benign compared to the devilish little demons of yore who stole children, played tricks on householders, and seduced innocent Christian knights. These contemporary fairies also developed wings—a borrowing from the earlier Victorian fairy revival. Real faeries never had wings. The new fairies lived in trees and flowers, were helpful to human beings, and had kindly pet unicorns—an animal that was not necessarily benign in the Middle Ages. They collected fairy houses and created fairy gardens and even began holding fairy house festivals, one of which a few years ago I happened upon.

The historic park at Spanish Point in Osprey, Florida, holds an annual fairy house festival each year, and since I was in the area at the time, so with nothing else to do that weekend, I decided to attend.

The grounds and gardens of the park were crowded with all manner of little imaginary beings. There were a few leprechaun boys around, and I happened to see Robin Hood there, but the attendees were mostly winged fairy girls in colorful little gowns and tutus and ribbons and insect antennae headbands. There was a park map indicating the location of little fairy villages with their tiny houses and gardens, and there were prizes given for the best ones. There was also music and storytelling, and there were many fairy godmothers in silky gowns and flowery headbands, all of them kindly—no demonic fairy queens, nor any demon lovers.

Along with the revival of fairy lore there is also, as it turns out, a darker side to this return to belief in unseen worlds and normally invisible beings. Forty-five percent of Americans currently believe in demons and ghosts, and there are still people who are afraid of vampires as well as revenants. In fact, some fourteen percent of Americans believe that zombies are going to rise up and start hunting for human flesh.

Along with these various festivals and beliefs in the invisible world that so intrigued Cotton Mather, there has also been a return to some of the old pagan religions. Included in this general movement is a reinterpretation of that anathema of the early Christian church—the witch. The belief in witches as evil doers appears early on in Judeo-Christian religions.

The Old Testament was quite clear on the matter of witches: they must be put to death. But in the pre-Christian era, older women, or crones, were believed to harbor knowledge of sacred and healing herbs and potions and had access to the spirit world of nature. Powerful wise-women, capable of good or evil deeds, occur throughout early European cultures. Some of these figures were mythological characters, such as Baba Yagar of Russian folklore who dwelt in the forest in association with wildlife and could dispense evil or good. But in most pre-literate societies around the world there were shamans, medicine women, witches, and herbalists who could cure diseases and even act as therapists who could diagnose problems through dreams or drug-induced journeys into the spirit world.

Now, as a result of research by contemporary feminist scholars, the positive side of witchery has reemerged. Many of the plants used by female shamans, or *shamanistas*, were more efficacious than the hideous and bizarre practices of traditional doctors of the past, practices such as bleeding patients or employing the doctrine of signatures, in which plants such as liverworts that resembled internal body parts were employed as curative medicines.

Ethnobotanists working with traditional healers among societies in Africa, the Americas, and the Indies have verified the science behind the traditional uses of plants. Shamans are able to identify hundreds of species of local plants and have wide knowledge of their various uses in human illnesses. Among the Mazatec people of northern Oaxaca in Mexico, for example, the curanderos have an intimate relationship with the local vegetation and its uses. There are many cultural groups in the region who use the local plants, but the Mazatecs are said to have the ability to understand the language of plants and can hold conversations with them. They even give certain useful plants personal names and address them formally in their ritual chants and prayers. As it must have been with Tom Doublet and the Christianized native people of the Beaver Brook watershed, conversion did not necessarily exclude local traditions. The Catholicized Mazatecs address some of their healing plants by saints' names, "San Pedro," for example, or "San Isidro," and ask them to help in their rituals.

Western researchers working with local healers have now determined that this seemingly magical use of plants has been proven to have healing effects for various diseases, including cancer. As outsiders, gaining knowledge of the secrets of these traditionally sacred plants is not always simple. I used to know a botanist from the University of Chicago who made the acquaintance of a very old Quiche Mayan curandero in western Belize and asked him to teach her the uses of some of his plants and how to apply them. Basically, he said he couldn't do that. So she didn't ask again, but she stayed on in the village for a year, working alongside the people in the cornfields. Finally, the old man indicated that he would help. Shortly thereafter he died, leaving a Western-educated stranger as the only one in the group who knew the healing plants and their usage.

Modern drug companies have analyzed the chemical contents of these traditional plants and have managed to synthe-
size the chemicals and reproduce them as modern commercial
drugs, often sold for exceedingly high prices. But many ethnobotanists and the anthropologists who have spent years gaining
the confidence of native shamans in order to learn their local
medicinal practices are concerned about the commercializa-
tion of their findings. Although they appreciate the development of the new beneficial drugs, they have a problem with
what they see as the exploitation of the shaman's knowledge
and have been working to get drug companies to financially
reward their sources.

· · ·

This renaissance of traditional festivals and beliefs has even
reached into the old Puritan towns in the Beaver Brook watershed, and also the towns of the Concord and Sudbury River
watersheds. A version of an ancient spring festival takes place
during the spring equinox and the summer and winter solstices
in Concord. And in a variation of the old European May Day
celebration, the local Concord art center holds its annual procession of papier-maché animals, accompanied by drumming
and chanting to celebrate the return of spring, the river gods,
and the Earth and all its beings.

Coupled with this rebirth of traditional festivals, there has
also been a political shift that seems to attract people on both
the left and the right, as well as those with no interest what-
soever in politics. This movement, if you could call it that,
involves people who are giving up on the modern world and, in
the manner of Henry Thoreau, are attempting to simplify their
lives. Wall Street brokers and financiers, doctors, businessmen,
and especially, it seems, computer geeks and engineers are giving up the sweet smell of financial success and are downsizing

and, whenever possible, attempting to live off the grid. Some move to small urban apartments, some move out to the country and try to grow their own food, and some of the more adventurous throw all caution to the wind—literally—pack up their families and put to sea.

This cultural shift is basically a reconsideration of the whole grand experiment that began with the industrial revolution and is perhaps understandable. We live in a time of social, economic, and ecological unraveling, capped by a future in which the coastal cities of the world will flood and millions of environmental refugees will be forced to move to drier ground. Even as the coastal lands dry up and islands drown, the human population is expected to increase by one billion in the next fifteen years. And all this is coupled with the massive extinctions of plants and animals in what is currently termed the Anthropocene, a veritable geological age similar to the Pleistocene in which climate and extinctions and environmental changes are caused not by natural changes but by human activity.

Our way of living seems to be passing into history, along with the principles that once framed our civilization—the myth of continuous progress and the belief of human domination of nature.

The end of the world as we know it may not be the end of the Earth: life will go on. But it is, perhaps, the end of civilization.

Thoreau summed this up on the slopes of Mount Katahdin when he was first faced with the inhuman wilderness of the primal earth. "Who are we?" he wrote. "Where are we?"

• • •

In keeping with the local return of festivals, I decided to stop at three different local events that took place around the winter solstice, each with a distinctive character. The first was basically a Christmas cocktail party. There were bright lights and a lot of drinking, laughter, banter, and noise. And also jazz. Musicians from the region stopped in and jammed with the hosts.

The second event was a solstice celebration. This one turned out to be a rather solemn gathering of people collected in a circle around a low burning fire. People were encouraged to take up a handful of tobacco in the Native American tradition and, if they were so moved, step forward with an invocation, or a wish, a prayer, a memory, or whatever they wished to express. Once having spoken they were instructed to throw the handful of tobacco into the fire.

This was essentially a borrowing from some Native American ceremony, although from what I know of traditional Native America gatherings from Kata there would have been harangues at such an event delivered to encourage warriors to take arms against an enemy just before an attack, or a circling shuffling dance, accompanied by monotonous drumming and high whining vocables—a demonic ritual, according to the Puritan colonials who had witnessed such events.

In this case, the parishioners left the fire circle in silence, a somber line of people leaving a dying fire.

The third festival was a traditional solstice bonfire celebration out of an old forgotten European culture. The huge bonfire had been high and nearly unapproachable before I arrived, but by the time I got there it had burned down to a few flames. Someone in the group had a tambourine and the people gathered around the fire were dancing or singing and circling the flames. Periodically, people would step back, rush the fire, and leap over it. This went on for an hour, sometimes with couples leaping hand in hand and then spinning off in a dance. No one was taking the event seriously. There was wine and laughter, and some people had dogs and paraded them around the fire, and soon enough others followed, leaping and snorting like pigs, howling and barking, and the dogs began to prance, and women began ululating like Bedouins or wild Indians, and slowly, ember by ember, the fire died and people went inside to eat a late dinner.

By midnight the party was over.

Some social critics have a little problem with what they call cultural appropriation—adopting the rites of various traditional cultures, such as those of the native peoples of the Americas, and practicing them within the powerful commercial, capitalist culture of the United States. One would not want to examine this criticism too closely. Yoga, tai chi, meditation, and similar ancient beneficial practices would have to be eliminated to be true to traditional American culture. Furthermore, most of the practitioners of these alien cultural practices would argue that they are honoring the culture of other societies.

Nevertheless, one would not want to go too far into a few of the ancient traditional rites, some of which involve nasty practices such as self-flagellation, scarification, and the hardships of initiation and coming-of-age ceremonies.

To be politically correct, I suppose, only the drunken Christmas cocktail party that I attended would be legitimate, according to the criticism.

• • •

On Christmas Day that same year, I went down to my place beside the brook before dawn—the sacred period known as "first light" among the local Wampanoag people. There had been a light snow the day before, but I cleared a dry spot and watched as the black, as yet unfrozen waters slipped between the still standing lion-brown stalks of the cattails. The sky was lighting in the east, but the sun was still below the ridge and the wall of trees on the eastern bank made a black curtain for an as yet to be realized day or, by extension, an unknown year to come.

Things were falling apart in the world that year. There were no fewer than eight wars raging at the time, and there were uprisings and outright rebellions. Nation was pitted against nation. There were food riots in some places, starvation in others, floods, mudslides, massive forest fires, assassinations, and rising dictatorships. Not one nation was at that time overly concerned about the coming cataclysm of global climate change, even though those in the scientific community were predicting doom.

World events notwithstanding, as I watched, the cold winter-white sun slowly rose above the wooded ridge to the east. The black waters slipped past the bank just as they have for the last twelve thousand years, and in the midst of that overarching silence, just as the sun cleared the tops of the trees on the opposite shore, I heard the long, spirited twittering of a winter wren.

The sun is but a morning star, Henry Thoreau wrote at the end of *Walden* as a way of putting things into perspective. The same could be said of the running waters of Beaver Brook. Cultures may come and cultures go; myths and legends may flourish and fade, but the stream goes on forever.